# The Concept of Peace in Judaism, Christianity and Islam

# Key Concepts in Interreligious Discourses

Edited by
Georges Tamer

In Cooperation with
Katja Thörner

**Volume 8**

# The Concept of Peace in Judaism, Christianity and Islam

Edited by Georges Tamer

DE GRUYTER

ISBN 978-3-11-068193-2
e-ISBN (PDF) 978-3-11-068202-1
e-ISBN (EPUB) 978-3-11-068222-9
ISSN 2513-1117

**Library of Congress Control Number: 2020946097**

**Bibliographic information published by the Deutsche Nationalbibliothek**
The Deutsche Nationalbibliothek lists this publication in the Deutsche Nationalbibliografie; detailed bibliographic data are available on the Internet at http://dnb.dnb.de.


Typesetting: Integra Software Services Pvt. Ltd.
Printing and binding: CPI books GmbH, Leck

www.degruyter.com

# Preface

The present volume in the book series "Key Concepts in Interreligious Discourses" (KCID) contains the results of a conference on the concept of peace in Judaism, Christianity and Islam held at the Friedrich-Alexander-University Erlangen-Nuremberg. The conference, which was organized by the Research Unit "Key Concepts in Interreligious Discourse" with the greatly appreciated support of the Evangelical Church in Germany (EKD), took place in Erlangen on December 14–15, 2017.

The Research Unit KCID offers an innovative approach for studying the development of the three interconnected religions: Judaism, Christianity and Islam. With this aim in mind, KCID analyzes the history of ideas in each of these three religions, always considering the tradition of interreligious exchange and appropriation of these very ideas. In doing so, KCID investigates the foundations of religious thought, thereby establishing an "archaeology of religious knowledge" in order to make manifest certain commonalities and differences between the three religions via dialogic study of their conceptual history. Thus, KCID intends to contribute to an intensive academic engagement with interreligious discourses in order to uncover mutually intelligible theoretical foundations and increase understanding between these different religious communities in the here and now. Moreover, KCID aims to highlight how each religion's self-understanding can contribute to mutual understanding and peace between the three religious communities in the world.

In order to explore key concepts in Judaism, Christianity and Islam, KCID organizes conferences individually dedicated to specific concepts. A renowned set of researchers from various disciplines explore these concepts from the viewpoints of each of the three religions. The results of each conference are published in a volume appearing in the abovementioned book series. Particularly salient selections from each volume are made available online in Arabic, English and German.

In this fashion, the Research Unit KCID fulfills its aspirations not only by reflecting on central religious ideas amongst a small group of academic specialists, but also by disseminating such ideas in a way that will appeal to the broader public. Academic research that puts itself at the service of society is vital in order to counteract powerful contemporary trends towards a form of segregation rooted in ignorance and to strengthen mutual respect and acceptance amongst religions. Such a result is guaranteed due to the methodology deployed by the research unit, namely the dialogic investigation of the history of concepts, as documented in the present volume on the concept of peace.

I wish to thank Dr. Albrecht Döhnert, Dr. Sophie Wagenhofer and their assistants at the publisher house De Gruyter for their competent caretaking of this

https://doi.org/10.1515/9783110682021-202

volume and the entire book series. I would also like to thank Mr. Ezra Tzfadya for his assistance in preparing the volume.

Georges Tamer
Erlangen, May 2020

# Contents

**Preface — V**

Alick Isaacs
**The Concept of Peace in Judaism — 1**

Volker Stümke
**The Concept of Peace in Christianity — 45**

Asma Afsaruddin
**The Concept of Peace in Islam — 99**

Georges Tamer, Katja Thörner and Wenzel M. Widenka
**Epilogue — 159**

**List of Contributors and Editors — 169**

**Index of Persons — 171**

**Index of Subjects — 175**

Alick Isaacs

# The Concept of Peace in Judaism

## A Vessel That Holds a Blessing

## Prologue

Since 2009, I have been engaged in a project based in Israel called Siach Shalom (Talking Peace).[1] Working on this project has meant embarking on a deep journey into the meaning of the powerful, complex and elusive concept of *shalom* in Jewish thought. Siach Shalom is essentially an effort to discover the secret of peace by turning the conversation about it into a practice which seeks to achieve it. My colleagues and I place the idea of seeking to discover the meaning of peace at the heart of the dialogue groups that we facilitate between religious and secular Israelis; Israelis and Palestinians. I have learned so much from this journey that I cannot dare to write about this topic without first acknowledging the debt that I owe to Siach Shalom and all of the participants in our dialogue groups. Most of all, I must mention my two partners in this work: Prof. Avinoam Rosenak[2] and Ms. Sharon Leshem Zinger,[3] from whom I have learned the most. Writing anything on this topic without accrediting them would be a scholarly crime. In mentioning them by name I hope to fulfill the Talmudic precept captured in the phrase, "Rabbi Elazar said that Rabbi Hanina said: Whoever reports a saying in the name of he who said it brings redemption to the world" (Babylonian Talmud Megillah 15a). If there is

1 Siach Shalom (Talking Peace) is a non-partisan civil society peace project that was co-founded by Prof. Avinoam Rosenak, Ms. Sharon Leshem-Zinger and Dr. Alick Isaacs in 2009. Siach Shalom operates under the aegis of Mishkenot Sha'ananim in Jerusalem. The problem our work aims to address is the mishandling of religion and the deep internal schisms this has created in both political processes and NGO interventions in the regional peace process. In this latter sense, Siach Shalom is also devoted to building cohesion and internal understanding inside Israeli society.

2 Avinoam Rosenak is a professor of Jewish thought and Jewish Education at the Hebrew University in Jerusalem.

3 Sharon Leshem-Zinger is one of Israel's leading group dynamic facilitators and psychodramatists who in addition to her work in Siach Shalom has taught at many places including Ben Gurion University and Sapir College (where she founded the Collot BaNegev group dynamic facilitation training program).

https://doi.org/10.1515/9783110682021-001

anything in this paper that agrees with their teachings, I learned it from them. If any of it strays, the responsibility is mine.

# 1 Introduction – Two Meanings of Peace in Jewish Thought

Peace is not an undiscovered subject in modern Jewish scholarship. A great deal has been written about the Jewish ideal of peace and the different ways of attaining it.[4] It seems quite obvious that contemporary interest in this topic is at least in part due to the unfortunate fact that the Jewish State has been embroiled in conflict since the day of its inception. Having survived without a pronounced political identity for thousands of years and after returning to the stage of international politics, the Jewish collective has found the legitimacy and the security of its identity challenged militarily, politically and ethically by a chronic state of political conflict quite unlike anything that Jews have experienced in history. While many have been driven by this reality to look beyond the Jewish tradition, for example to the progressive values of the west, to find their answers; there is indeed a very significant effort to seek peace inside the teachings of Judaism and the number of initiatives, research projects, books and essays that this has yielded is indeed a blessing that has made much of the Torah's teaching about peace readily available to all who seek it.[5]

Given this, I think it is important in this paper to try to present something a little different. Rather than repeating what has already been written, I think it would be more valuable to investigate the religious history of the particular meaning of peace that in my view is most relevant to the contemporary Middle East but which is most overlooked in scholarship. This is a way of thinking

4 I would like to thank Rabbi Dr. Daniel Roth for his extensive work in this field and for the bibliographical material he has provided me with. See for example Gopin, Marc, *Between Eden and Armageddon*, Oxford: Oxford University Press, 2000, 167–195; Kaminsky, Howard Gary, *Traditional Jewish Perspectives on Peace and Interpersonal Conflict Resolution*, New York: Teachers College Columbia University, 2005; Steinberg, Gerald M., "Jewish Sources on Conflict Management Realism and Human Nature," in: Michal Roness (ed.), *Conflict and Conflict Management in Jewish Sources*, 10–23, Ramat Gan: Bar Ilan University Program on Conflict Management and Negotiation, 2008; Roness, *Conflict and Conflict Management in Jewish Sources*, 140–141.

5 See Kaminsky ibid. for a detailed bibliography and summary of the field especially 30–34.

about peace that many associate with the most dissenting religious voices on the Jewish side of the conflict and as such it is often disregarded or even vehemently opposed. Since I don't want to address this topic in sociological or political terms and I certainly don't want to identify my position with that of any particular political group, I think it might be useful to begin by offering a philosophical distinction between two fundamentally different dimensions of peace in Jewish thought. These two are not the only meanings of the word *shalom*, but the use of a binary distinction here serves the purpose of clarity and gives me a point of entry into the analysis that follows.

The first dimension frames the meaning of peace quite conventionally in the religious values and practices that Jews turn to when they seek to resolve situations of conflict. There are indeed many examples in Jewish thought and in Jewish law of peacemaking practices that come to resolve arguments, disagreements and even violent conflicts that erupt between individuals,[6] families, communities, peoples – Jews and non-Jews. The Jewish tradition is very rich in legalism and the idea that a legal system or a judge can be an arbitrator in a situation of conflict is not foreign to the *halakha* (Jewish law) by any means. Similarly, throughout Jewish history we have examples of peacemakers and dispute resolvers who, emulating the great biblical example of Aaron the Priest, sought to resolve differences between conflicting parties without resorting to the judgment of the courts.[7] Bearing in mind some of the more recent terminology developed in the field of conflict resolution, it is possible to find traditional Jewish examples of resolving, managing and transforming conflict as well as practices that we might readily compare with alternative dispute resolution (ADR). This dimension of peace and the classical texts associated with it is the one that has attracted the bulk of scholarly attention in the field and it is not the one that I wish to address in any further detail in this paper.

In counter–distinction to the more conventional examples of peacemaking found in the Jewish tradition, the second dimension of peace refers specifically to the unique conditions that apply to the end of days and the messianic redemption. This peace is the ultimate world peace that the prophets spoke of and which is associated in the Bible with the ingathering of the exiles to the land of Israel, the return of the entire land to the Jewish people and the fulfillment of the biblical covenant. This form of peace, which I have previously

---

**6** Kaminsky, ibid. Part IV, 190–218.

**7** Roth, Daniel, *The Tradition of Aaron Pursuer of Peace between People as a Rabbinic Model of Reconciliation*, PhD diss., Ramat Gan: Bar Ilan University, 2012.

referred to elsewhere as both "Prophetic" and "Messianic" peace,[8] is often considered an obstacle to the resolution of conflict. The notion that something of messianic proportions is taking place in the 'here and now' can easily be used as a foil for resisting the more practical work of negotiation, compromise and agreement that Realpolitik demands. This observation is not without justification. However, since the notion of prophetic peace is the one most concerned with the conditions that many religious Jews in Israel understand as taking place in the world today – i.e. the return of Jewish exiles to the biblical land of Israel – I submit that clarifying the irenic potentiality of this concept is the more relevant and meaningful challenge to tackle at this time.

## 1.1 Prophetic Peace and the Ingathering of the Exiles

Prophetic peace in Jewish thought is a concept that is fundamentally connected to the fulfillment of the Jewish purpose in history. It is a form of peace that is grounded in a theological ideal that includes more than just the cessation of a particular military conflict. It is in fact the resolution of all internal and external conflict in the human soul, in intimate relations, in the family, the community, the Jewish people, international politics, nature and indeed between human beings and God. As it appears in the Bible, this kind of peace brings with it a total transformation of human consciousness and of the conditions of human personal, social and political life as we know them. This is the peace that the prophets speak of, that biblical teachings are geared towards and that the prayers that observant Jews recite every day yearn for. It is a meaning of peace that is more closely connected to the Hebrew word '*shalom*' (from the Hebrew root Shin, Lamed, Mem – meaning wholeness and completion) than the English word 'Peace' (from the Latin Pax – meaning pact or agreement).

The objection that holding out for completion runs the risk of obstructing more immediate and practical solutions to present-day problems is valid. The notion that the higher dream prevents people from taking certain steps towards lesser but more realistic achievable goals is one that needs to be taken very seriously. This is especially true if these steps can directly improve a pressing situation or alleviate human suffering. All the same, my suggestion is that widespread belief in prophetic peace is a concept that we cannot ignore. It is also a kind of peace that we can work with as we endeavor to create understanding between

8 Isaacs, Alick, *A Prophetic Peace. Judaism, Religion and Politics*, Bloomington: Indiana University Press, 2011.

people and address the complex and painful conflict that has surrounded the State of Israel since its establishment in 1948. This is true both because the vision of prophetic peace is by far the most central principle of peace in Jewish thought and because the vision of prophetic fulfillment is a powerful force in contemporary Israeli religious Jewish identity. This is a vision that is built upon a great deal of ancient wisdom that has much to teach us today. This vision, therefore, is both authentic to the mainstream of classical Jewish thought and relevant to the contemporary situation.

For many "national religious" Jews living in Israel today, the conflict in the Middle East is not an isolated or detached modern experience. Rather, it can be seen as a crucial stage in the very long journey that the Jewish People has been on for thousands of years. This journey begins with the Jewish religious obligation to fulfill its collective covenantal purpose as outlined in the Torah.[9] That purpose is one given in covenant to Abraham, Isaac and Jacob; to the tribes of Israel and to the people who emerged from bondage in Egypt and who then stood together to receive it prophetically at Sinai. The purpose of this covenant is to be a holy people united in a holy land where they are to be a blessing, as God says to Abraham in Gen. 12, to all the families of the world. The covenant of Sinai insists that through living the life prescribed by the Torah, the Jewish people united in the land of Israel will disclose the unity of God to the world. Disclosing a consciousness of God's unity is likened in numerous Jewish sources to the shining of a light and it is perhaps most famous in the writings of Isaiah who spoke prophetically about the day when the Jewish people will become a light unto all the nations of the earth.[10] As many national religious Jews see it, the main story of our present period in history concerns the fulfillment of this covenant. After thousands of years of exile, the people are finally returning to the land and rebuilding it. But, their struggle to return and to re-form their collective identity is one that has been plagued by conflict and political opposition. For many, this opposition is a spiritual event which has deep meanings many of which are not known or understood, but which guide Israel toward the fulfillment of its prophetic purpose. These are meanings that need to be uncovered in order for the lessons of recent history to lead us in the direction of unity and peace. For them, this vision is very real and practical and its obstruction by conventional, political and diplomatic peacemaking practices is something that stirs up vehement spiritual, Halakhic and political opposition. Appreciating this

---

**9** See Soloveitchik, Joseph Dov (1903–1993), *Kol Dodi Dofek* (*Fate and Destiny. From the Holocaust to the State of Israel*), New York: Ktav Publishing House, 2000, 42–44.
**10** Isa. 49:6.

is crucial to understanding the widespread opposition of the religious communities in Israel to diplomatic peace efforts in the last 30 years.

If we put this idea in slightly different terms, we might say that for many religious Jews, it is no accident that the conflict in the Middle East seems to defy the capabilities of modern diplomacy. It is spiritually and religiously significant that the framework for peacemaking that modern politics provides is emerging as inadequate to the task of imagining a workable solution to this situation. And so, it seems valuable, and perhaps even essential, to try to think beyond the limits of secular politics and consider the possibility that the working definition of peace that conventional diplomatic practices of peacemaking are based upon is not appropriate to the task at hand. If the Jewish narrative of return to the land is indeed a step toward the fulfillment of the biblical covenant, then it seems reasonable to imagine that the failures of western diplomacy in the region are grounds enough to turn to the prophetic concept of peace and see what we can learn from it.

## 1.2 The Three Elements of Prophetic Peace

Having said a few words about the authenticity and relevance of our topic, in what follows I will try to explain the meaning of "Prophetic Peace" as my colleagues and I have come to understand it. Prophetic Peace is a complex idea, and I therefore want to present it systematically by dividing it up into three component parts. Though these three elements can often appear separately in Jewish texts, my claim is that they coincide significantly in the full concept of prophetic or messianic peace. Thus, I submit that a deeper understanding of each one and, most particularly, of the connections between them, is the key to unpacking the meaning of shalom in Jewish thought.

The three elements of prophetic peace are:

1. Anti-Politics
2. Unity of Opposites[11]
3. Knowledge of God

11 This concept has been developed most significantly in the research of Avinoam Rosenak who has dealt with its central role in the teachings of Rabbi Abraham Isaac Kook as well as in the extensive sources in Jewish thought upon which Rav Kook draws. See for example Rosenak, Avinoam. "Hidden Diaries and New Discoveries. The Life and Thought of Rabbi A.I. Kook," *Shofar. An Interdisciplinary Journal of Jewish Studies* 25:3 (2007), 111–47; *Prophetic Halakha. The Philosophy of Halakha in the Teaching of Rav Kook*, Jerusalem: The Magnes Press, 2007, 44–56 [Hebrew].

I will first introduce the concept of anti-politics giving illustrative examples of how it has appeared in biblical and rabbinic texts. I will then pick up the theme of the unity of opposites presenting examples of its biblical and rabbinic history. Next I will trace the connections between these and the knowledge of God showing how the combination between the three can offer us a definition of prophetic peace that we will be able to see in modern religious texts. Finally, I will offer some insights and suggestions, gleaned from the work of Siach Shalom, into ways we can think about the practical value of Prophetic Peace in the context of today's conflict in the Middle East.

## 2 Anti-Politics

'Anti-politics' is not strictly speaking a "Jewish" term but it is useful for our purposes because it characterizes several concepts that are central to the Jewish understanding of God and the collective. George Konrad used the phrase "anti-politics"[12] in a book of that name that some would argue helped bring down the Soviet regime in Central Europe. Konrad urged his readers to think of "anti-politics" as a realistic way of dealing with political oppression. His book *Anti-Politics* argued *for* standing down and *against* engaging in confrontation. Konrad proposed a notion of: "*de-statification*", which basically meant imagining a political system characterized by a reduction of power from above. Ultimately anti-political thought seeks to protect society from the volatile fusion of a grand idea with political power.

Though this was not Konrad's intention, his phrase is very useful for describing a profound element of the prophetic ideal in which the vision of peace is connected to a feature of Jewish religious thought that downplays the role of power in the life of the collective. In religious Jewish thought, the nation of Israel is not a political community of individuals held together by a common origin or government. Rather the Jewish collective is primarily understood as an expression or even as a creation of the uniting will of God, which brings the people together through their shared obligation to collectively live the life prescribed by the Torah. Rather than applying force or building a lowest common denominator around which groups can rally, the Torah is addressed to the ideal of a People who can only serve God together. In order to unite in this way the People must align their individual and collective will with his will as an act of free-choice. Thus the national community is a full expression of the freedom

12 Konrad, George, *Anti-Politics. An Essay*, trans. Richard E. Allen, New York: Quartet, 1984.

of each individual who finds his or her own unique place in the collective observance of the Torah by freely choosing it. This freedom depends on what the Torah refers to as "*hester panim*" i.e. the concealing of God's face.[13] This is the concept that makes space for people to choose rather than being too heavily imposed upon by the divine presence. Similarly, the Kabalistic tradition emphasized the notion that free-choice and even the basic independent existence of the world are only made possible by God's withdrawal or constriction of his light (i.e. of our awareness of him) in the world. Kabbalistic texts refer to this idea as *sod ha-tsimtsum* which literally means 'the secret of [God's] constriction'[14]. Both of these ideas, *hester panim* and *tsimtsum*, underline the principle that freedom or room for choice is made possible by – what is perhaps the ultimate anti-political act of – self-effacement and withdrawal from power. In the context of this withdrawal, the notion that divine sovereignty or *malkhut shamayim* and covenant or *brit* has an anti-political nature emerges into view.

From the prophetic perspective, peace has no obvious place inside the individualistic, power-laden and belligerent political process at all. The prophetic notion of peace is not about conventional political action. On the contrary, the biblical visions of peace seem to suggest that an ideal peace for Israel can never be the direct outcome of political action at all but must rather emanate from a "circumcision of the heart"[15]. This inner transformation (which is the cumulative outcome of all the free choices that observance of the Torah requires Jews to make every day) is described by the biblical prophets as something that happens when the Jewish people return from exile to collective life in the holy land.[16] The phrase "circumcision of the heart" is a metaphor for the removal of a hard covering that prevents the heart (meaning the inner consciousness) from recognizing God and his perpetual presence in (and as) creation. The removal of this covering demands a profound psychological shift in how human beings interact with one another, with the world and with God. In this context peace is achieved through a kind of anti-political politics in which power is replaced by listening; negotiation by spiritual engagement; interests-based agreements and alliances by genuine efforts to live together in a loving unity that mirrors or echoes the true depths of human consciousness in a place where it merges with a total awareness of God.

13 Deut. 31:17.

14 Tzimtzum is a term widely used in Lurianic Kabala. A useful explanation of the term in its various forms can be found in Kaplan, Aryeh, *Inner Space*, New York: Moznaim Publishing, 1990, especially 120–128.

15 Deut. 30:1–6.

16 Ibid.

## 2.1 Anti-Politics in the Bible

The anti-political theme occurs in the Bible both in terms of biblical theology and with reference to the organization of the collective life. Naturally, these two strands also overlap. The anti-political ideal of human life perhaps appears first and etiologically in the infamous choice presented to Adam and Eve in the Garden of Eden.[17] Despite the long association of their choice with sin, the question remains, what is so wrong with the fruit of a tree that gives knowledge of good and evil? Why should God forbid human beings from knowing the difference between good and evil? Why should the Bible choose something so noble as the knowledge of good and evil as the object of the serpent's temptation?

Rabbi Samson Raphael Hirsch among many other traditional Jewish scholars understands this choice as an opportunity for Adam and Eve to resist the temptation of animalistic self-assertion.[18] It is this temptation that is associated with the animal or bodily side of the human self and as such it is the intuitive choice of the serpent. As animals, humans and serpents alike are instinctively anxious about self-preservation for which the cunning to distinguish between good and evil is indeed essential. The choice to eat of the fruit of this tree is therefore the choice between the natural human instinct for self-preservation, self assertion and perhaps even self- redemption and the choice to overcome these instincts in an act of restraint.[19] This restraint opens the door to the possibility of freely embracing the consciousness associated with the divine soul or *neshama*. In this framing, the fruit of the Tree of Knowledge represents the presumption of humans that they can manage the world without God. From this tree humankind acquires the ability and the need to build a robust human political technological society.[20] However, as spiritual beings endowed with the divine soul which exists within and which is constantly enraptured in communion with God, choosing the fruit of this tree means turning away from a higher or inner anti-political option. Resisting the fruit of the tree would have meant yielding the urges of the ego to an act of self-restraint. As Rabbi Kook explains, *sod hagevura* or the secret of self-restraint means cultivating a point of contact

**17** Gen. 2:16–17.

**18** Hirsch, Samson Rapheal, *The Pentateuch. With Translation and Commentary*, New York: Judaica Press, 1962. Reissued in a new translation as Haberman, Daniel, *The Hirsch Chumash*, New York: Feldheim/Judaica Press, 2009; Gen. 2:16.

**19** See Rabbi Abraham Kook, *Lights of Holiness*, vol. 3, Jerusalem: Mosad HaRav Kook, 1985, 199 (Hebrew).

**20** Weinreb, Friedrich, *Roots of the Bible. An Ancient View for a New Outlook*, trans. N. Keus, Braunton: Merlin Books, 1986, 183–220.

with deeper or higher desires than those which serve selfhood and ego.[21] Restraint at the end of the sixth day would have meant relinquishing human power in exchange for entry into a world without power represented eternally by the Sabbath day – the anti-political day – in which work, commerce and anxiety over the conquest of space are suspended.[22]

Many classical Jewish commentaries see the choice between individualizing self-assertion and finding one's place in unity with God as a root structure that recurs throughout the Bible. This is the choice that we are referring to as the dilemma between political and anti-political action. The contrast appears in the story of Cain and Abel in which Abel relies upon God by offering to him the best of his flock while Cain, resourcefully keeping the best for the needs of his own survival, offers God less.[23] For this he is sent away from the land and made ironically dependent on God for the rest of his life. The choice between life in the presence of the divine and the power of knowledge appears again in the structured contrast between the covenantal line of Shem and the people of Shinar who built the Tower of Babel.[24] Indeed, the destruction – or perhaps deconstruction[25]– of the city and tower, provides us with one of the clearest anti-political metaphors in the Bible. The etiological story of Babel – in which the people band together in a city to protect themselves from their anxious fear of being scattered across the face of the earth – is an explicit example of biblical reticence about the polis. The people of Shinar seek to live in a political world of their own making and to ensure its permanence by building a tower to the heavens. Though the biblical text offers no clear indication that this desire is an act of defiance against God, classical rabbinic interpretation (Midrash) sees it in this way[26] understanding the construction of the city as a choice to live in a man-made world rather than in God's world. Similarly, the tower is seen as a means to enter the heavens and declare war on God.[27] This interpretive tradition, though not literal, is underlining the binary structure of the choice. The route that the people of Shinar take follows their own ingenuity and creative

**21** Kook, *Lights of Holiness*, 199.

**22** Heschel, Rabbi Abraham J., "Prologue: Architecture of Time," in: *The Sabbath*, 2–10, New York: Farrar Strauss and Giroux, 1951.

**23** See Gen. 4 and Weinreb, *Roots of the Bible*, 221–236.

**24** Gen. 11.

**25** Jacques Derrida, "Des tours de Babel," in: *Difference in Translation*, ed. and trans. Joseph P. Graham, Ithaca: Cornell University Press, 1987, 165–248.

**26** See Talmud Sanhedrin 109 a and Targum Yerushalmi (Yonatan) Gen. 11:4.

**27** "Let us build a city," He will come down to us and *we will ascend to heaven* and if not, *we will declare war on Him.* Despite this He left them alone and said to them "do as you will." See Midrash Tanhuma, Gen. Section 18.

power along a path that leads to their independence from God. This independence is echoed in the biblical detailing of how the city was built from man-made bricks that the people manufacture for this purpose.[28] The tower itself is perhaps then a synthetic extension of the Tree of Knowledge because it is built with the human technology that the fruit of that tree revealed. The polis is therefore understood as a godless place that human beings can only imagine living in because of the Tree. As such the story of the city and politics in general can be seen as the human journey away from the source, away from the Sabbath, away from the Garden and away from God. It is a rejection of the journey back to the Oneness of the divine and hence it leads to the very dispersion that the people of Shinar most feared.[29]

In direct contrast to this, the Bible tells the story of the line of Shem. The Hebrew letters Shin and Mem that spell the word Shem occur time and again in the Babel story meaning "name", or "there". This recurring leitmotif in the story prepares the reader for the subsequent passage and builds the contrast between the dispersion of Babel and "These are the generations of Shem."[30] The line of Shem represents the Tree of Life; it is the line of Abraham, Isaac and Jacob. These are the bearers of the covenant who create the curve in history that eventually leads back to the as yet uncelebrated Sabbath day in the Garden of Eden. Abraham follows God almost blindly to a place that he is only told he will be "shown."[31] The famous words of God's commandment to Abraham, "*lekh lekha*" (literally, 'go unto yourself') suggest that his journey to Israel is a journey inwards toward the deeper self. Even on the surface of the text it is clear that Abraham does not know where he has been commanded to go and why he must go there. He only knows that he must follow the path to God and hence to the prophetic calling that he hears or perhaps visualizes in his inner self. Thus, Abraham is only told what he must leave behind: his land, his birthplace and his father's house.

In the land, Abraham lives outside the city. He is a shepherd whose existence in the land is presented in anti-political contrast to that of his nephew Lot who chooses to inhabit the cities of the Jordan Plains. The text knowingly tells us that these are like the land of Egypt and they include the cities of Sodom

**28** See Gen. 11:4.

**29** See Weinreb, *Roots of the Bible*, 278–282.

**30** Gen. 11:10.

**31** In Gen. 12 God commands Abraham to travel to the land "that I will show you." The Hebrew word for "show" has the same root as the word Moriah as in Mount Moriah or the Temple Mount in Jerusalem.

and Gomorrah that will soon be destroyed.[32] These cities are God forsaken places where morality has collapsed and human society has failed. Despite Abraham's best efforts to save his kin, the temptation of the city overwhelms the wife of Lot who cannot restrain herself from yearningly looking back at Sodom and Gomorrah even as they go up in flames behind her.[33] The collapse in Lot's moral conduct which is most blatantly obvious in his willingness to send out his daughters to be raped by the people of the city is then monumentalized in the destiny of his offspring who he begat incestuously and unknowingly when his daughters take sexual advantage of him in return.[34]

The anti-political motif recurs in the repeated passing of the covenant from father to younger (rather than first born) son. Abraham is chosen over Lot's father Haran; Isaac over Ishmael; Jacob over Esau and Joseph over Reuben. The theme continues when David, the youngest of Jesse's sons is anointed King and followed by Solomon who is King in place of David's first born, Absalom. There is an anti-political metaphor in the vision of a bush that is not burned by fire, in the miracles with which Moses fails to convince Pharaoh and in the triumph of slaves over the Egyptian Empire. All of these images surrounding the leadership of Moses suggest that biblical prophecy is perhaps the opposite of the Greek *parrhesia*. Anti-politics is not speaking truth to power, it is not the politics of the powerless. It is the choice of spiritually motivated self-restraint against the urge to seize power. It is the recognition that the Ma'apilim, the band of hotheads who sought to conquer the land of Canaan after God had pronounced his punishment for the sin of the 10 spies, learned the hard way when they were slaughtered by the Amalekites.[35]

By this stage it should be clear why the characteristics that we are calling anti-political are inherently connected to the prophetic genre. Prophecy with its vertical (the prophet and God) and horizontal (the prophet and the people) dimensions promotes a spiritually driven social order that seeks to establish unity and oneness in spiritual communion with God rather than through the use of human charisma or power. This idea, which emanates in the withdrawal and restraint of God, is ideally echoed and mimicked in the life of God's people, Israel.

On the vertical axis, this is the very nature of divine sovereignty itself which emerges in the quietist images of God that abound in both the Torah and in subsequent biblical passages. God is subject to repeated disobedience because he conceals his face from the people throughout history and is to be

32 Gen. 13:10–11.
33 Gen. 19:26.
34 Gen. 19:30–38.
35 Num. 14:40–45.

found, as Elijah discovered on Mount Horeb, in total utter silence.[36] God is not in the thunder, the fire or the blustering wind.[37] The jealous angry and punishing God of western Old Testament scholarship is not the biblical God as understood in Jewish traditional exegesis but rather an 'aspect' of revelation called 'Din' or Judgment that results from the lack of internal peace and alignment between the people and God. When people are true to their inner selves this jealous aspect of God called *'Din'* and the merciful aspect called *'Raḥamim'* which are One are finally experienced by humanity as One.[38] Thus the ideal of divine sovereignty itself emerges as the gentlest of things. Its anti-political structure represents the possibility of peace and reconciliation between God and humanity that is understood as humanity's return to its truest inner self and to the journey back from the path that Adam and Eve took when they eat from the Tree of Knowledge. Anti-politics leads back to the Tree of Life. This is the meaning of the idea that the Torah itself is a Tree of Life,[39] a new covenant that is ingrained in the inner consciousness of the people and inscribed in their hearts[40] "whose ways are gentleness and whose paths are peace."[41]

On the horizontal axis, the prophet seems devoid of political appetite. Jeremiah says, "I cannot speak for I am a child"[42], Amos says, "I was no prophet, neither was I a prophet's son; but I was a herdsman and a dresser of sycamore trees."[43] Moses said, "O Lord, I am not a man of words."[44] The prophet is slow of speech, self-consciously unconvincing, politically disarmed. He seems ill- suited to the charismatic rhetoric we associate with human power and in this the prophet is a place-marker for the gentle revelation of God. Thus God says to Jeremiah, "Say not: I am a child. I am with you. I shall forever protect your speech, and whatsoever I shall command thee thou shall speak."[45] The Lord touches Jeremiah's mouth.[46] He feeds a honey-sweet scroll into the mouth of Ezekiel;[47] the angel scorches the mouth of Isaiah with a burning coal to purify

**36** 1Kings 19:11–12.
**37** Ibid.
**38** See Midrash Rabbah on Gen. 12:14, Yalkut Shimoni on Gen. 19.
**39** Prov. 3:17.
**40** Jer. 31:30–34.
**41** Prov. 3:17.
**42** Jer. 1:6.
**43** Amos 7:10.
**44** Exod. 4:10.
**45** Jer. 1:6.
**46** Jer. 1:9.
**47** Ezek. 3:3.

his speech.[48] The prophet needs no speechwriters; he needs not agonize over the appealing lucidity of his phrase making. And he needs not expect to be heard. Thus the prophet stammers and stutters, speaks reluctantly and in the wonderful phrase coined by A. J. Heschel, in a tone that is an "octave too high".[49] Even the revelation of the Torah at Sinai is given to a people who have first chosen it and who cannot manage to hear it on their own.[50] All of these (and many more) are anti-political images associated with the genre of prophecy.

'Anti-politics' in the Bible is therefore a two tiered idea. On the first level, the resistance to power-based political solutions to human problems is contrasted with divine sovereignty. A classic example of this is the passage in Deuteronomy in which Moses recounts the commandment to establish a King in the Promised Land.[51] The king must be modest; he must refrain from owning too many horses or from marrying too many wives. But, most importantly, his kinghood is subservient to God. His role in society is a fulfillment of a divine command which in its essence is no different from that of a Cohen, a Levite, or any other man or woman; all of whom play the role they do in the collective in accordance with the commandments of the Torah. The particular expression of this that reminds the king of his status is the commandment to scribe a Torah scroll of his own which not only binds him to the law like a constitutional monarch but binds the very concept of the monarchy to the collective divine service in which he plays only his part.[52] When he violates his role or misuses his power, not only is the Kingdom stripped away, but the entire hierarchical power system of the city in which he rules is reduced to nothing. As the prophet Isaiah writes,

> For behold, the Lord, the Lord of host doth take away from Jerusalem and from Judah stay and staff, every stay of bread and every stay of water. The mighty man and the man of war, the judge and the prophet and the diviner and the elder, the captain of fifty and the man of rank, and the counselor and the cunning charmer and the skillful enchanter and I will give children to be their princes and babes shall rule over them.[53]

On the second level, the anti-political motif applies to God and is ingrained in the very DNA of prophecy. The prophets obsess over the injustices that the

**48** Isa. 6:7.
**49** Heschel, Abraham J., *The Prophets*, New York: Harper and Row, 1962, 10–12.
**50** Sommer, Benjamin, "Revelation at Sinai in the Hebrew Bible and in Jewish Theology," *The Journal of Religion* 79:3 (1999), 422–51.
**51** Deut. 17:14–20.
**52** Ibid. 17:17.
**53** Isa. 3:1–5.

powerless endure and seem out of touch with political expediency. Their message – which is to bear witness to the possibility of revelation – cannot be heard and is rarely heeded. It gives no power to the prophet and instead takes it away. It is perhaps this sting that Jonah the son of Amitai (meaning 'the truthful') could not endure. Perhaps this is why he ran away from the task given to him by God. He worried that if his warnings were heeded God's mercy and forgiveness would undermine his reputation as a speaker of truth. He feared the anti-political nature of the prophetic calling. Perhaps we might say that his story describes how the temptation to follow the Tree of distinction between good and evil brought him to seek the just destruction of Israel's enemies in Nineveh rather than their forgiveness. In this book the messages are not then ones that the prophet preaches but those that he learns himself. It is his prophetic calling to veer away from the path only then to be shown the way back to the Tree of Life by God himself.

## 2.2 Anti-Politics in Rabbinic Literature

There is much controversy in Jewish scholarship about the relationship between the Bible and the rabbinic tradition. In this paper, which deals expressly with the meaning of peace in the spiritual realm of religious Jewish thought, I will follow the traditional path that seeks to ground rabbinic hermeneutics in the multi-layered reading of biblical teaching rather than seeing it as an evolution or perhaps even a deviation from it. Critical historical scholarship seeks out evolutions, twists and turns, departures and inconsistencies and often accounts for them by politicizing them; offering up 'external influences', 'expediency' and 'alternate agendas' as explanations for all of these. In this framing, the rabbinic tradition with its emphasis on law and authority is political or legal in its nature and, according to some scholars, even devoid of theologically motivated religious passion. It is a legal system with an evolving moral conscience reacting hermeneutically to the challenges that historical events pose to the maintenance and development of the biblical tradition. In this sense it is seen as a legal/political system that seeks to adapt biblical law and perhaps wield biblical authority to its own ends (for better and worse) and in service of its own moral and social agenda, in an evolving and changing reality.[54]

---

**54** This is an analysis of the relationship between divine revelation and rabbinic interpretation that is famously associated with the work of David Hartman, see in particular Hartman, David, *The Living Covenant*, Woodstock: Jewish Lights Press, 1980; see also Fisch, Menachem, *Rational Rabbis. Science and Talmudic Cultur*, Bloomington: Indiana University Press, 1997; Halbertal,

Conscious as I am of these scholarly arguments, I feel that they are not helpful here. At their very core they follow a presumed chain of power that the tradition constructs in either good or bad faith in order to maintain a system in which rabbis can monopolize religious authority previously held by priests and prophets. From my perspective, this highly documented and learned scholarly approach, championed by many great experts in the field, fails to access the inner code, the DNA of the rabbinic ideal which in my humble view, like the prophetic one, is anti-political in its nature.

In order to understand this, we need to explain the close integration of oral and written Torah as the tradition understands it. The concept of the "oral law" which many scholars see as a late development in Jewish thought can also be understood as inherent to the Bible itself. The biblical text, composed in Hebrew letters but without the vowel signs that are required for correct pronunciation, cannot be read definitively without an oral tradition that punctuates and vowels it. That tradition, though not without rigidity, is also endowed with an inner flexibility that is attested to by the precedents set by biblical characters themselves whose stories are part of the Torah. In rabbinic literature, this flexibility also expresses itself in what might appear to those unfamiliar with rabbinic hermeneutics as a kind of manipulative wordplay. In Hebrew, words themselves are open to and indeed invite multiple readings. The interactions and interrelations between words, with and without standardized vowel signs, endow the original Hebrew text of the Bible with a sort of fluidity and multi-vocality which rabbinic hermeneutics can augment by employing what the rabbis refer to as the 13 principles of Torah exegesis. These include such methods as importing the meanings of words in one context and then applying them to a quite different one. Similarly, certain combinations of letters are regarded as interchangeable. As such the fabric of the text, the morphology, the numerology, the etymology, the semantics and even the pronunciation of words, shift in and out of meanings that cannot be reduced or simplified. Thus, the language of rabbinic literature is endowed with the anti-political stammering of the prophets.

Our suggestion is that following the destruction of Jerusalem, rabbinic literature pursues prophecy by other means seeking to construct a detailed regimen of practice geared toward the ultimate purpose of preparing the Jewish people to be the vessel of revelation and light that they failed to become in biblical times. However, the rabbinic ideal of prophecy is not one of authoritarianism

Moshe, *People of the Book. Canon, Meaning and Authority*, Cambridge: Harvard University Press, 1997; Sagi, Avi, *The Open Canon. On the Meaning of Halakhic Discourse*, trans. Batya Stein, London and New York: Continuum, 2007.

but rather of anti-politics. In this sense, the relationship between what the rabbis referred to as the oral and the written laws is one of total interdependence. If the written law, the book of the Torah, is likened to a projector; the oral law, the life of the people, is a screen. Neither one can fulfill its purpose without the other. The Torah requires the people and the people require the Torah. As such the balance between them is not authoritarian but inherently alive. The oral law, or the Halakha, is better understood when we compare it to a law of nature than it is when compared to a law of the State. The rabbis are spiritual guides in the ways of this law much more than they are legislators and senators who wield its power.

An example of this rabbinic ethos is captured in the following discussion about the transfer of authority from prophets to rabbis that presumably followed the destruction of the second Temple in Jerusalem by the Romans in the year 70. The Talmud (Baba Batra 12a) states,

> R. Abdimi from Haifa said: Since the day when the Temple was destroyed, prophecy has been taken from the prophets and given to the wise. Is then a wise man not also a prophet? What he meant was this: Although it has been taken from the prophets, it has not been taken from the wise. Amemar said: A wise man is even superior to a prophet, as it says, "Ve´navi Levav Ḥokhma."[55]

The point that the Talmud is making in this passage requires some teasing out. The primary thrust of the original statement seems to suggest that authority has been transferred from the Prophets to the rabbis following the destruction of the Temple. However, if we reject the notion that prophecy is authoritarian we can uncover the concern of the text more accurately. The text immediately questions the assumption that prophecy has been transferred from prophets to rabbis raising the logical objection that if prophecy has been transferred in this way then surely we should simply treat the rabbis as prophets from now on? Since this is not logical, the text seeks to make a clearer distinction between rabbis and prophets. It suggests that they each reflect a different form of prophecy which I propose are best understood as two different forms of closeness to God. Thus the passage reads that while one form of closeness to God ended with the destruction of the Temple, the other form, the rabbinic form, continues to exist.

The nature of this second form of prophecy, I wish to suggest, is as anti-political as that of the first even though the two forms of prophecy are significantly different from one another. The key to understanding the second form is hidden in the biblical passage that Amemar cites to support his seemingly radical claim that "the wise man is superior to the prophet". The proof text, cited

55 Talmud Bavli Baba Batra 12a.

from Psalms 90 is understood in the context of a rabbinic wordplay to mean that “wisdom is the heart of prophecy” and hence superior to it. However, to any learned student of the Bible and the Talmud it is clear that this interpretation is not a literal one. The more literal reading of the biblical verse understands the word “*ve´navi*” which can mean “and a prophet” to mean “and we will bring”. Thus the verse should be read, “And we will bring wisdom to the heart” and not, as Amemar’s reading suggests “and wisdom is the heart of prophecy.”[56] The two meanings of the word “*ve´navi*” are interchanged here by the rabbis in order for a key principle to emerge in a slightly ironic way.

This key principle states that closeness to God and understanding his will can be accomplished via rabbinic interpretation. Thus rabbinic interpretation of the biblical text is seen as a prophetic (or a prophecy gaining) enterprise. In rabbinic “prophecy” the method of gaining closeness to God is connected more clearly to a method of reading than it is to any single conclusion drawn from the text. The method elicits and plays with the duality of the word “Ve´navi” which, due to its morphological multiplicity, attains meaning in a non- authoritarian way. Thus the rabbinic method elicits deeper, hidden or simply additional meanings from the biblical text by allowing the words to resonate with a kind of multi-vocality that gives the student access to the many meanings of Torah. This multiplicity is what the rabbis identify with the complexity of prophetic experience because the many layers of the text are seen to correspond with the many different levels of human consciousness which combined correspond to the different layers of divine revelation in the soul. The complexity of the human soul that can receive revelation is thus echoed in the layers of the Torah that bestow it. In its very nature rabbinic interpretation is an anti-political spiritual practice that is the very opposite of authoritarian, monolithic, dogmatic legalistic or political rhetoric.

The idea that this anti-political prophetic ethos is perhaps a founding element of the rabbinic tradition is exemplified in the story that the Talmud relates concerning Rabbi Yohanan Ben Zakkai at the time of the Temple’s destruction in Jerusalem.[57]

> Abba Sikra, the head of the bandits of Jerusalem, was the son of Rabbi Yohanan Ben Zakkai’s sister. Rabban Yohanan sent to him saying, “Come to me secretly.” He came. Rabban Yohanan asked him, “How long are you going to carry on this way and kill all the

56 This wordplay is made possible by the double meaning of the Hebrew word Navi which can mean “prophet” and also is the first person plural future of the verb to bring i.e. “we will bring.”
57 Talmud Gittin 56 a–b.

> people with starvation?" He said to him, "What can I do? If I say a word to them, they will kill me." Rabban Yohanan said to him, "Devise some plan for me to get out of the city perhaps I can save a little."

In the plan that unfolded, Rabbi Yohanan fakes his death in order to get out of the city. When he exists from the city walls, he greets the highest officer of the Roman legions in what appears to be an inappropriate way.

> When Rabbi Yohanan came to Vespasian he said, "Peace to you O king." Vespasian said to him, "you have been condemned to death on two counts, firstly because I am not a king . . . and secondly if I am a king why did you not come to me until now?" Rabbi Yohanan said to him, "In truth you are a king, for Jerusalem can only be destroyed by a king as it is written, And the Lebanon shall fall by a mighty one (Isa 10:34) And 'mighty one' means only a King as it is written, 'And their mighty one shall be of themselves and their ruler shall proceed from the midst of them' (Jer. 30:21) indicating that 'mighty one' parallels 'ruler'. And Lebanon means only the Temple as it is stated, 'That good mountain and the Lebanon (Deut. 3:25) and for your second point that I didn't come to you until now, there are zealots in the city who would not allow it . . . '. At this point the messenger arrived saying, 'Arise for the emperor is dead and the notables of Rome have decided to make you head of the state.' Vespasian was overjoyed and he said to Rabban Yohanan . . . 'You may make a request of me and I will grant it.' Rabban Yohanan said give me Yavne and its wise men . . . He ought to have said to him, 'Let Jerusalem alone.' But Rabban Yohanan thought that Vespasian would not grant so much.

This is a complex story with many implications that need not concern us here. Our focus is on two things. First, this story is crucial to the founding ethos of Rabbinic Judaism in the time following the destruction of the Temple. Though it is not to be taken literally as historical account, it is clearly a foundation narrative[58] that establishes the leadership of Rabbi Yohanan Ben Zakkai on the anti-political principle that rejected the struggle of the zealots or "bandits" against the Romans and sought to salvage Torah study as an anti-political practice that poses no threat to Roman rule. This is expressed in his willingness to make peace with Vespasian without seeking power. He does not ask to save Jerusalem or even the Temple but only to build a Bet Midrash in Yavne where the practice of Torah study can continue.

Second, the prophetic nature of Rabbi Yohanan's interchange with Vespasian is presented here as something that is deeply entrenched in the rabbinic method. His interpretations of the biblical verses are Midrashic. In this case they exemplify the exegetic methodology mentioned before, of drawing the meaning of words in

---

**58** On the importance of foundation narratives to the ethos of Yavne see Fisch, *Rational Rabbis*, 51–55.

one context and applying them to another. Thus 'mighty one' and 'rule' are equated with 'king'. Next, 'Lebanon' is equated with the 'Jerusalem Temple' in order to yield the final prophetic insight that Jerusalem can only be conquered by a King. Ultimately, the story reassures the followers of Rabbi Yohanan that Midrashic exegesis endows rabbinic leaders with prophetic insight. Given that this situation is also one of confrontation, it is not insignificant – though not essential to our overall theme – that Rabbi Yohanan emerges as a peacemaker. What is essential to our overall theme is the connection between anti-politics, this method of exegesis, and the understanding of peace or *shalom* as a "unity of opposites".

## 3 The Unity of Opposites

The concept of 'the unity of opposites' is the one most obviously associated in classical Jewish thought with *shalom*[59] or prophetic peace. Peace is the coexistence of conflicting points of view as in the words of the great Hassidic master, Rabbi Nachman of Breslav who wrote, "Peace is the unity of two opposites".[60] What is important to point out here is the inherently anti-political nature of this concept. Not only is this framing of peace paradoxical in nature and thus more similar to the stammering contradictions of biblical prophecy than it is to straight shooting political rhetoric; but the possibility of this paradox is ultimately enabled by the same anti-political theology of *hester panim* that we discussed above.

The unity of opposites is a term that describes the specific nature or architecture of God's unity. As with anti-politics, our contention is that the unity of opposites is also part of the very DNA of Jewish thought. It is a feature that is present not only as an idea or a value. Rather it is a concept that shows us the heart of how biblical and rabbinic texts function since they notoriously contain different and sometimes contradictory accounts of the same events and laws. Perhaps the most obvious examples are the two creation stories that appear in chapters 1 and 2 of Genesis and the two versions of the Ten Commandments that appear in Exodus and Deuteronomy. Similarly, rabbinic literature abounds

59 As mentioned above, Avinoam Rosenak has played a central role in my understanding of this principle especially with reference to its centrality in the teachings of Rabbi Kook. His own research is accredited in the notes below, but many of the sources quoted in this section have also been made available to me by him.
60 Rabbi Nachman of Brestlav, *Likutei Moharan*, Teaching 80.

in contradictions and disagreements which traditional Jewish thought perceives of as a unified system of thought. The paradoxical nature of these textual structures exemplifies the prominence and centrality of the unity of opposites and hence of peace, to Jewish thinking.

The core idea here is that when the bible teaches that "God is 'One' and his name is One" it is essentially pointing to the notion that God is the Being that unifies everything in diversity. Put differently, God is the infinite Oneness that emanates into the infinite diversity of creation. In this sense, as Rabbi Yehuda Ashlag – the great twentieth century Kabbalist – taught, "'*ein od milvado*" i.e. everything is God and there is nothing other than God.[61] The reason why the world does not appear to us in this way is *hester panim*, the act of concealment in which God allows the world to be itself, as it were. The prophetic message of the Torah does not teach us to see God so much as it teaches us to see this world as his creation i.e. as an emanating expression of his being. The return to unity at the culmination of history (i.e. prophetic or messianic peace) is therefore not a step away from multiplicity or this-worldliness into a realm of metaphysical unity but rather an expansion or ultimate extension of this worldly multiplicity into a uniting this-worldly conscious awareness of this world as a unified divine creation. This paradoxical ideal, which insists on the co-existence of two conflicting and mutually exclusive realities, symbolizes the prophetic vision of peace. Images of paradoxical radical co-existence abound in biblical, rabbinic and later texts. This is perhaps captured most famously in the prophecy of Isaiah in which he envisions a time when the "wolf will lie down with the lamb."[62] The metaphor here is clearly one of peace between two opposing forces but it is one that cannot be achieved through negotiation or the aligning of political interests but rather through a process of spiritual transformation after which we view the world as a paradoxical union of unity and diversity which is inherently anti-political in nature.

If we return to the anti-political nature of Jewish theology as exemplified by the principle of *hester panim*; the concealment of the divine consciousness from humans is also the fabric upon which the world is created in all its complexity. If we consider creation as an explicit expression of God's will, the non-prophetic or everyday human experience of the world is one in which that deeper reality is concealed. The singularity of the divine which is expressed in the world "as the world" is accessible to us only apophatically i.e. in everything

---

**61** This idea is based on two verses in Deut. 4:35 and 4:39. See Sutton, Avraham, *Spiritual Technology*, New York: Hebrewbooks, 2013, 34–39.
**62** Isa. 11:6.

that we experience as not being God. What we in fact see when we look at the world is an infinite variety of conflicting individualities all made possible by the anti-political withdrawal of God from consciousness. The journey back to closeness with God is therefore a journey to a perspective or a point of view about the world in which we can see all of this individuality and conflict as a unity of opposites that emerges from singularity and then culminates in it. This is a formula that is captured in the numerology of the Hebrew word for 'Father' which reads 1-2-1.[63] In other words, the path to peace is one that requires us to recognize the unity that is made in our father/creator God's image (1) that emerges into the world as the duality of creation – light and dark land and sea etc. (2) and then ultimately in messianic peace returns to One (1). To seek this oneness in multiplicity is to pursue prophetic peace in God's image. By way of contrast, negotiating agreements and making compromises – the business of political peace – generates a form of sameness that is made in our own human image. In this sense the unity of opposites is an anti-political concept that is made possible by the anti-political principles of *hester panim* and *tsimtsum* and which calls human attention to the underlying oneness of the created world.

## 3.1 Unity of Opposites in the Bible

In biblical theology, the unity of the opposites is at the heart of the biblical idea that God is One. Again, when the Bible is read as the tradition understands it, the different names of God as well as the many different contradictions that characterize the revelation of God are all expressions of a unity that is God. Hence the transcendental Elohim who 'speaks the world into being' in the first chapter of Genesis is a different revelation of the same inherent form of being "*havaya*" that moves immanently in the Garden of Eden interacting with creation and with humankind. This unity is expressed in many places but perhaps most explicitly in Exod. 6:2,

> And Elohim spoke to Moses saying I am Havaya

This explicit unity of these two names is at the heart of what must be considered the most basic and generally recognizable statement of Jewish thought. This is, of course, the "*sh'ma*"; the biblical phrase from Deut. 4:1, "Hear Israel the Lord Our God the Lord is One". This verse which is uttered endlessly in Jewish prayer might be better understood when rendered, "Hear Israel Havaya

---

**63** See Weinreb, *Roots of the Bible.*

our Elohim Havaya is One." In the words of Arthur Green, "this cry that stands at the center of our worship . . . is a call to all who struggle with the divine and the human, who struggle to understand." He continues,

> This is the higher unity, the inner gate of oneness. According to the unity of Sh'ma all is one as though there were no many. Nothing but the One exists. God after Creation and God before Creation are one and the same . . . The world makes no difference. Its existence is totally unreal or totally inconsequential from the point of view of the One . . . . The garbing of divine energy in the countless forms of existence is naught when seen from the point of view of infinity. Only the One is real.[64]

When this verse is recited in prayer it is followed by the words "Blessed is the name of God's glorious kingdom forever and ever". Green goes on to comment on the meaning of this saying,

> This is the lower unity, the outer gate, the one within the many. We refer to it as the unity of God's kingdom. Here we encounter God's oneness in and through the world, not despite it. Each flower, each blade of grass, each human soul, is a new manifestation of divinity, a new unfolding of the cosmic One that ever reveals itself through its multicolored garments, in each moment taking on new and ever-changing forms of life. In the variety of life's riches we discover the unity that flows through them all, the divine life that animates all of being.[65]

In other words, the juxtaposition of the two statements, the first of which is uttered purposefully out loud while the second is whispered, captures the duality embodied in the unity of opposites. The former represents the higher One, the latter represents the lower expression of this One in the form of a two and together they are a prayer for the return to the One. The crucial principle that Green does not mention here is that this journey from the total singularity of the divine that remains unaffected by creation to the duality and multiplicity of creation is mimicked or echoed in the total unity of the people of Israel that stood together at Sinai "as one person with one heart"[66] and which lived in the land as one nation in one land. The second phrase is whispered because it echoes a lower condition in which that unity is concealed. In biblical metaphor this concealment is synonymous with the copula that connects sin with the dispersion of the Jewish people and their exile from the land. Their vertical separation from God is thus equivalent to their horizontal separation from the land. This structure of paralleling distance from God with other images is echoed

**64** Green, Arthur, *Seek My Face. A Jewish Mystical Theology*, Woodstock: Jewish Lights Publishing, 2012, 4f.
**65** Ibid.
**66** See the commentary of Rashi to Exod. 19:2.

along many other axes. For example this distance can also be seen to represent the separation of the written Torah from the oral Torah since the Sh'ma is biblical and the second phrase is rabbinic. On Yom Kippur they are recited together in full voice as a symbol of the special unified condition that the Jewish People attain on the holiest day of the year. It is a joyous day of atonement and redemption from sin not a sad day of mourning. Yom Kippur is a Sabbath among all the Sabbaths of the year thus it is a taste of the ultimate Sabbath that was lost in Eden when the sweet taste of the fruit was withdrawn from bark of the tree[67] and the world was given over to distinctions between good and evil. The unity of opposites is the reuniting of good and evil, of the fruit and the bark, the hearts of fathers and their children,[68] God and creation, absolute Oneness and multiplicity, holiness and profanity, men and women, right and left and – perhaps most controversially today – the dispersed people and the divided land. In the unity of opposites, all of the divisions and dualities that we experience in the world become explicit as extensions of the Unity of God and thus in the metaphor of Ezek. 37:15 the tree or the stick of Judah and the tree or stick of Joseph will become One in the hand of the prophet. When this happens, the people will return to the land and a covenant of peace will be made between them and God under the leadership of King David.

## 3.2 Unity of Opposites in Rabbinic Literature

Perhaps in counter-distinction to the biblical vision of the unity of opposites, the rabbinic one is more focused on the proliferation of multiplicity. This is not an unbridled pluralism or relativism since the credentials required for participating in the expansion, interpretation and legislation of the oral law are demanding. All the same, this idea of multiplying the teachings of the oral law comes to bring the students of Torah closer to God by making the Torah greater. Making the Torah greater extends the Torah toward God. By means of apotheosis, the proliferation of multiplicity points more completely towards a higher or deeper unity that earthly diversity mirrors. This is often achieved by disagreement or as we have already seen, by the midrashic layering of the biblical text's meaning. This is the significance of Rabbi Yohanan Ben Zakkai's method which endowed prophetic significance to biblical verses by saturating their literal meanings with additional ones. This simple scene which describes the foundation of the Bet

67 See Yalkut Shimoni 8,1 and Rashi commentary on Gen. 1:8.
68 See Mal. 3:24.

Midrash (study house) in Yavne epitomizes the ideal of Torah study as it came to be seen in the time of exile.

Rabbinic interpretation seeks to maintain the possibility of closeness to God in an era in which biblical prophecy has ended. In this sense, it is, as we have said, prophecy by other means. It is a spiritual quest for God in a time of exile. While prophetic texts contend with the inner exile of the soul that is rooted in the expulsion from Eden, the worship of the Golden Calf and the evil report of the ten spies etc.; rabbinic literature contends with a further collapse – the political exile from Jerusalem, the destruction of the Temple, the cessation of prophetic revelation, the canonization of the now sealed biblical text and the dispersion of Jews outside of the land. It refers equally to the exile of Shehkina i.e. the lost access to divine consciousness, the hardening of the heart and the further concealment of God's face. In this time, the rabbinic tradition proposes a rigorous approach to Torah study which involves the close literal, allegorical, philosophical and mystical interpretation of the biblical text. The rabbis codified a complex regimen of Halakhic observance based on both the literal and allegorical interpretation of biblical law adherence to which was considered integral to the successful study of the text itself. They established new institutions of Jewish life such as the House of Study and the Court that came to replace the land, the monarchy and the sacrificial ritual of the priests. If, as we have suggested, all of these are anti-political in their ideal form (this is perhaps the secret of Jewish survival in exile), the rabbinic tradition is founded on the ideal of perfecting them for a future time in which they will accomplish their goal of uniting the inner being of human consciousness with divine will while emerging from the condition of exile and achieving the messianic ideal of prophetic peace.

A striking Talmudic passage illustrates the centrality of this ethos to the rabbis while underlying the co-existence of anti-politics and the unity of opposites in rabbinic thought,

> R. Aha bar Hanina said: It is revealed and known before Him, Who spoke and the world came into being, that there was none in the generation of R. Meir like him; why then did they not fix the Halakha according to his view? Because his colleagues could never fathom the depths of his reasoning. For he would assert that something unclean was clean and make it seem plausible, and he would assert that something clean was unclean and make it seem plausible. A Boraitha taught: his name was not actually R. Meir rather Nehorai was his name. Why then was he called Meir? Because he made the eyes of the sages shine in the law. And similarly, Nehorai was not his name rather R. Nehemiah and others say R. Eliezer ben Arach. Then why was he called Nehorai? Because he lit up the eyes of the Sages in the Law.[69]

**69** Talmud, Eruvin 13b.

This passage presents us with a surprising portrayal of one of the greatest scholars of rabbinic history, Rabbi Meir. Rabbi Aha bar Hanina begins by equating the general appreciation of Rabbi Meir's brilliance with what is revealed before – or known to – God. This expression of what 'God knows' is especially significant in this context because it is loaded with double entendre. On the one hand, Rabbi Aha bar Hanina seems to be saying that Rabbi Meir's genius is common knowledge. At the same time, since 'what is known to God' is perhaps known only to God, he is suggesting that there is something hidden about Rabbi Meir's Torah. This paradoxical duality recurs throughout the text. We are told that Rabbi Meir is the greatest teacher of his generation but that the law does not follow his rulings. Beyond the obviously anti-political nature of this which is indeed of great significance to our overall argument; the reason given here why his teachings do not determine the law is that they are so contradictory or paradoxical that no one can completely understand them. In other words, anti-politics and the unity of opposites coincide here. Rabbi Meir is introduced to us as a master whose teachings attain what we might describe as a 'prophetic' outcome which, as the first sentence of the passage suggests, is both revealed and concealed from us.

Rabbi Meir exemplifies the anti-political nature of rabbinic thought. The greatest scholar is the one whose teachings are not followed. However – and this is the whole point here – this is not because his legal rulings are wrong, but because an unequivocal legal outcome is perceived as an inadequate condensation of a deeper complexity which can never be captured in one point of view alone. In God's world, law is perceived as a unity in which right and wrong, purity and impurity, the sacred and profane even truth and falsehood are too implicated in and connected to one another to ever be completely separated. The unequivocal voice of political legislation is therefore not one that can capture the truth about the law as Rabbi Meir sees it. And hence, in a political world, the law cannot possibly follow the paradoxical complexity of his readings. However, this does not mean that his teachings cannot be identified by the Jewish legal system – a system that has survived a long history without political power – as representing an ideal in which the co-existence of conflicting points of view is more important than the choice or decision that yields a single outcome. This is a point of view that is made possible by the unique nature of the Halakhic deliberative process which records debates about the law and gives more attention to these debates than it does to the final rulings. Indeed, as impractical as this diverse anti-political structure sounds, one cannot deny the virtues of a legal system that has successfully held a people together through thousands of years of unparalleled dispersion and adversity.

R. Meir's articulation of conflicting legal premises through ingenious halakhic argumentation has theological significance since in this literature the concept of the law is equivalent to the will of God. When he demonstrates his ability to generate endless paradoxical combinations of conflicting points of view through his mastery of Halakha he is essentially displaying his insight into the unity of opposites that characterizes the nature of divine will. His seemingly exoteric teachings intertwined to create an esoteric experience of revelation. His rulings cannot determine the practical outcome of legal disputation but they serve the higher purpose of elevating the study of the law to a level at which concealment coincides with revelation. This is the spiritual meaning of the law. It is the characteristic of the law that reflects the paradox of divine revelation and it is this quality in Rabbi Meir's insight that qualifies him as the greatest scholar of his generation.

Tellingly, this passage appears immediately before a highly significant and well known Talmudic story that describes the prolonged disagreements between the school of Bet Hillel and the school of Bet Shammai. The Talmud says,

> R. Abba stated in the name of Shmuel: For three years there was a dispute between Beth Shammai and Beth Hillel, the former asserting, "The halakha is in agreement with our views" and the latter contending, "The halakha is in agreement with our views" Then a heavenly voice (Bat Kol) issued announcing, "These and these are the words of the living God and the halakha shall follow the rulings of Beth Hillel." Since, however these and these are the words of the living God, why is the Halakha to follow the rulings of Beth Hillel? Because they were kindly and humble, they studied their own rulings and those of Beth Shammai and mentioned the teachings of Beth Shammai before their own.[70]

The most common scholarly readings of this text identify in it the basic Jewish rabbinic principle of pluralism.[71] In this conflict, two points of view compete unrelentingly until a divine voice is forced to intervene (this might be taken as a literary device employed by the text and not necessarily as a spiritual reality that the text expects us to take seriously). The divine voice then chooses between the two points of view stating that each has an equal truth claim while the position of Bet Hillel is the one that will determine the law because it is the more tolerant.

I want to highlight an alternative reading of this well known text which is based in the commentary of the famous Rabbi Judah Loew of Prague (known as "Maharal of Prague")[72] and Rabbi Kook but which is essentially built upon the

**70** Talmud Eruvin 13b.

**71** See Fisch, *Rational Rabbis* and Sagi, *The Open Canon.*

**72** Maharal, *Derekh Hachayim*, Bnei-Brak: Yahadut Publication, 1980, 258f.

insight of Sharon Leshem Zinger.[73] Leshem Zinger suggests that it is the debate between the rabbis that is responsible for generating the divine voice.[74] In other words it is their disputation that engenders prophetic insight and hence the ruling of the voice is not to be understood as a pluralist statement about the equal validity of the two conflicting points of view. Rather, as Rabbi Kook understands, it is a spiritual statement about the nature of the words of the living God. The words of the living God, or prophetic insight are the underlying peace that is expressed in this world as conflict and opposition. The unity of these two opposites is what the rabbis accomplished and it is this that the Maharal understands in the phrase "these and these are the words of the living God". For him this is a sort of equation which we might formulate as follows:

> These + these = words of the living God.
> Put slightly differently:
> Unity of opposites = prophetic peace.

This example offers us a profound way of understanding Talmudic and rabbinic disagreements which comprise a huge part of this entire literature. In certain places rabbinic literature acknowledges a sense of crisis about these disagreements and considers the possibility that they represent a lack of clarity about the law and by way of extension about the divine will. However, the attitude to disagreement in most cases is quite different. For the most part they represent a prophetic ideal. They are a lower-worldly expression of a deeper unity of opposites which mimics the paradoxical coexistence of ideas and even physical realities which, at a higher or deeper level of abstraction, are seen as a unity. This idea of higher abstraction is often represented by angels as in the following passage from Derekh Erets Zuta,

> Bar Kappara said: Great is Peace, as among the angels there is no animosity, no jealousy, no hatred, no commanding, no quarrelling, because the Holy One, blessed be He, has made peace among them, as it is written [Job, xxv. 2]: "Dominion and dread are with him: he maketh peace in his high places." "Dominion" is the angel Michael and "Dread" is

---

**73** Maharal, *Gevurot ha-shem*, Bnei-Brak: Yahadut Publication, 1980, 35; Neher, André, *The Teachings of Maharal*, Jerusalem: Reuben Mass, 2003 (Hebrew); Rosenak, Avinoam, "Unity of Opposites in the Teachings of Maharal. A Study of his Writings and Their Implications for Jewish Thought in the Twentieth and Twenty-first Centuries," in: Elchanan Reiner (ed.), *Akdamot*, 449–87, Jerusalem: Zalman Shazar Institute, 2015 (Hebrew). On R. Kook's doctrine of the unity of opposites see Rosenak, Avinoam, *Rabbi Kook*, Jerusalem: Zalman Shazar Center, 2006, 34–42 (Hebrew); *Prophetic Halakhah* 44–57 (Hebrew).

**74** Leshem-Zinger, Sharon, *The Vessel of Peace. A Cultural and Symbolic Model for Making Decisions Peacefully about Peace* (forthcoming in Hebrew).

> Gabriel, one of whom is of fire and the other one of water, and still they do not oppose each other, for the Holy One, blessed be He, has made peace between them . . . The name of the Holy One, blessed be He, is also "peace" (*shalom*), as it is written [Judges, 6, 24]: "And called it Adonay-*shalom*."

In this sense the unity of opposites which is perhaps the most common genre in rabbinic literature serves the purpose of granting angelic or prophetic insight into the unity of creation. God who is the underlying principle of Unity encompasses all opposites and as such is the ultimate peacemaker capable of uniting even fire and water. The idea that the coexistence of contradiction points toward the divine is also the theme of the following passage from Midrash Rabba,

> This is what the Torah states: Who will bring pure from impure, not one (Job 14:4). Like Abraham who was born of Terach, Chizkiah from Achaz, Yoshiah from Amon, Mordechai from Shim'I, Israel from idolaters, the world to come from this world. Who has done this? Who commanded this? Who ordered this? Is this not the work of the Only One? A spot on a man's skin is impure. If the rash spreads to his entire body he is pure. Who has done this? Who commanded this? Who ordered this? Is this not the work of the Only One? A woman whose child dies in her womb, if the midwife reaches in and touches the body she becomes impure for seven days while the mother is pure until the dead baby is delivered. If a man dies in a house the house is pure, when his body is removed it becomes impure. Who has done this? Who commanded this? Who ordered this? Is this not the work of the Only One? It was taught: those who are occupied with the red heifer from beginning to end make their clothes unclean. It itself (contrarily) purifies defiled garments. God said: I set a statute, I decreed a decree, and you cannot transgress my decree.[75]

This Midrash is essentially dealing with the theological significance of a legal category known as "*ḥok*". *Ḥok* which is generally translated as statute or statutory law is a commandment in the Torah that cannot be understood rationally. In this sense, the observance of *ḥok* is the quintessential expression of religious obedience to divine command. However, the rabbinic explanation of *ḥok* that is provided in this Midrash points to a deeper meaning that identifies inner contradiction as the essential quality that characterizes all of the laws in this category. The prime example of this inner contradiction is found in the biblical injunction to use the burnt ashes of a Red Heifer to achieve the ritual purity required for entering the Tabernacle or Temple. This legal passage in Num. 19, which is explicitly referred to in Biblical Hebrew as the *ḥok* of the Torah, stipulates paradoxically, that while the ashes of the Red Heifer purify those who

---

**75** Bemidbar Rabba 19,1 (my translation).

bathe in them, the person who prepares the ashes becomes impure for a period of seven days. Thus, "it itself purifies defiled garments" but those who are "occupied in the act of its preparation" become unclean. The conclusion that the Midrash draws from this and the other examples of *ḥok* is that paradox, contradiction and even conflict as in the case of Israel and idolaters, point to the underlying unity that underpins God's created world. "Who has done this? Is it not the One?" the Midrash concludes. All contradictory points of view in the law and in the world are seen as expressions of the same single unified will. The diversity and paradoxes in the Torah are all in the image of divine unity. This is a foundational principle of Jewish law as in the words of the Talmud[76],

> "The masters of assemblies": these are the disciples of the wise, who sit in manifold assemblies and occupy themselves with the Torah, some pronouncing unclean and others pronouncing clean, some prohibiting and others permitting, some disqualifying and others declaring fit. Should a man say: How in these circumstances shall I learn Torah? Therefore the text says: "All of them are given from one Shepherd". One God gave them; one leader uttered them from the mouth of the Lord of all creation, blessed be He; for it is written: "And God spoke all these words". Also do thou make thine ear like the hopper and get thee a perceptive heart to understand the words of those who pronounce unclean and the words of those who pronounce clean, the words of those who prohibit and the words of those who permit, the words of those who disqualify and the words of those who declare fit.

The singularity and the unity of the words of the divine shepherd represent a totality that contains the paradoxical coexistence of the contrary rulings of each of the rabbis. Those who rule that an object is pure and those who rule that the same object is impure both speak the words of the one shepherd. The transformation that needs to take place in order for this to "make sense" is also implied here. The teachings of the Talmud are both the ends and the means. They represent a unity of opposites and in this sense they are an end in themselves. However, the study of these disagreements opens the ear and the heart, creating the perspective that is required to see unity in the depths of what appears upon the surface to be a contradiction. It is this perceptive heart that ultimately recognizes in the words of both sides the words of the same shepherd. The conclusion here is of crucial importance: through the study of Torah human beings can learn to see peace and harmony where others see only contradiction and conflict.

76 Hagiga 3b.

# 4 The Knowledge of God

This capacity to see unity through the study of Torah is essentially a state of mind that the Jewish tradition refers to as the Knowledge of God or *da'at Hashem*. This state of mind is the third and final component of prophetic peace. In many ways it is the culmination of anti-politics and the unity of opposites in the sense that it refers to a state of consciousness in which the inner being of the self (*neshama*) is able to disclose to our regular consciousness (*nefesh*) that the world with all its paradoxes and contradictions is a unified created expression of God's wholeness (*shalom*). The Hebrew word for knowledge (*de'ah* or *da'at*) has an ontological as well as an epistemological meaning. The word suggests not only 'knowing' in the conventional sense but also the erotic kind that is often coyly referred to as '*knowing* in the biblical sense' – as in "and Adam *knew* Eve his wife"[77]. To know is to be mingled with or absorbed in just as fish and plants and other myriad life forms are absorbed in and united by the water. To know God, is to submerge one's consciousness in God. It is to know everything that is, as God. It is to attain a perception of all being as *being-in-God*. This is a condition of being that Maimonides in the twelfth chapter of his Laws of Kings describes as one in which there is no hunger, no war, no jealousy and no competition between people because the understanding that we are all different limbs in a single organism overtakes our sense of ourselves as individuals. This image, which is found in the teachings of the Maharal of Prague, Rabbi Kook, Martin Buber and many others, is especially powerful because it envisions a kind of unity between people that does not impose uniformity upon them. In the same way as the different limbs of the body must perform extremely different functions for the organism as a whole to thrive, every individual must somehow be him or herself in the fullest sense of their individuality in order for the overall unity that encompasses all forms of being, to be complete. This full inner consciousness is *da'at Hashem.*

The knowledge of God or *da'at Hashem* is intimate, penetrating and inward which is why it is associated not only with eroticism but with ingestion. The ingestion of the fruit of the Tree of Life is *da'at Hashem* and it stands in stark contrast to *da'at tov vera* – the ingestion of knowledge that separates good and evil – that conceals the intimate knowledge of God's unifying unity and focuses our attention on our naked animal self. In this sense, the tragedy of Eden is the loss of *da'at Hashem* and the purpose of Torah is to regain it.

77 Gen. 4:1.

This condition of knowing is one that requires humility, trust and restraint. It demands that we put our egoism aside not only when we think of our individual selves, but also when we imagine ourselves as members of groups. The regular concept of the collective sees the group as an expansion of the ego. The nation is an inflated self that Martin Buber famously compared to an idol.[78] The true nation, that nation that Abraham fathered is charged with the task of founding its collective identity on the biblical Covenant. It is the nation whose collective attention is absorbed beyond itself in a higher truth. Again, this absorption is the 'knowledge of God'. It is a state of being that propels the world toward unity and away from separateness. To paraphrase Buber's image, by living in the world as a united being, we can experience the unity of God in the world as the world and live in it in peace.[79]

## 4.1 Knowledge of God in the Bible

Many biblical passages address the centrality of this state of prophetic consciousness to the vision of unity and messianic peace in the Bible, but we will focus on one in particular which illustrates not only the meaning of *da'at Hashem* but also its connection to both anti-politics and the unity of opposites. This very famous passage appears in Isa. 11 in a prophecy that was mentioned above. The full passage reads as follows:

> And there shall come forth a shoot out of the stock of Jesse, and a twig shall grow forth out of his roots. And the spirit of the LORD shall rest upon him, the spirit of wisdom and understanding, the spirit of counsel and might, the spirit of knowledge and of the fear of the LORD. And his delight shall be in the fear of the LORD; and he shall not judge after the sight of his eyes, neither decide after the hearing of his ears; But with righteousness shall he judge the poor, and decide with equity for the meek of the land; and he shall smite the land with the rod of his mouth, and with the breath of his lips shall he slay the wicked. And righteousness shall be the girdle of his loins, and faithfulness the girdle of his reins. And the wolf shall dwell with the lamb, and the leopard shall lie down with the kid; and the calf and the young lion and the fatling together; and a little child shall lead them. And the cow and the bear shall feed; their young ones shall lie down together; and the lion shall eat straw like the ox. And the sucking child shall play on the hole of the asp, and the weaned child shall put his hand on the basilisk's den. They shall not hurt nor destroy in all My holy mountain; for the earth shall be full of the knowledge of the LORD, as the waters cover the sea.

78 Buber, Martin, "The Spirit of Israel and the World Today," in: idem. (ed.): *Israel and the World. Essays in a Time of Crisis*, 183–96, Syracuse: Syracuse University Press, 1984, 184.
79 Buber, in: idem., 184–87.

If we look carefully at the structure of this prophecy we can trace in it the three elements of prophetic peace. It begins with an anti-political image in which leadership is handed over to a child/shoot/twig. This child rules and judges the people; but not by virtue of his wisdom, experience or political strength. Rather, he makes his judgments and gives counsel without the power of his own eyes, ears or mind but with the knowledge of God. In this verse the knowledge of God emerges as closely connected to anti-politics in an image of a leader who is a child but who leads and judges all the same because it is not his power that he exercises but that of *da'at Hashem* – a higher consciousness that instructs him in the ways of unity, justice and peace. Next, in a reality that the interconnection of the knowledge of God and anti-politics creates, we encounter the unity of opposites in the images of the predators and prey lying together. Once again, since they are led by the anti-political image of a child two of our themes come together. Finally, we return to an image of the entire world living in peace on God's holy mountain. This peace is made possible by the knowledge of God that fills the earth 'as the water covers the sea'.

The image of water covering the sea is perhaps paradoxical and one that is difficult to understand. How can water cover sea when both are water? This paradox is precisely the point of the image and it mirrors the relationship between the light of a candle and the light of the sun. Both represent the paradoxical relationship between the individualistic consciousness of the self and the total consciousness of the divine. In the condition that the prophet is describing here, the individual's consciousness is most fully absorbed within its inner self when it is totally absorbed in its awareness of the divine. However, in this situation the self is not erased. Totality and individuality are united and thus this state of consciousness is in itself a unity of opposites or a peace between separateness and unity. Thus, anti-politics, the unity of opposites and the knowledge of God come together and enable each other in this image of world peace.

## 4.2 Knowledge of God in Rabbinic Literature

The ethos of attaining knowledge of God is central to the spiritual underpinnings that give religious meaning to the study of rabbinic literature. While this is indeed a central theme that appears widely and variously in this vast literature, it perhaps serves our direct purposes best to simply point out very briefly how this element of prophetic peace has been present in all of the rabbinic texts that we have cited above with reference to either anti-politics or the unity of opposites and then to move on to seeing some later rabbinic references to this idea in which its meaning is perhaps made more explicit.

*Da'at Hashem* appears in rabbinic literature as the spiritual or prophetic dimension of Torah study. It is essentially the rabbinic / prophetic outcome that we have seen attributed by the Talmud to the rabbis' teachings when these are considered superior to prophecy. We saw this in the encounter between Rabbi Yohanan and Vespasian and again in our understanding of the heavenly voice generated by the disagreement between Bet Hillel and Bet Shammai. We encountered it in the discerning ear that hears the words of the shepherd in the contradictory rulings of the rabbis and again in the state of mind that sees the unity of the One in the contradictions of *ḥok*. As we have seen, the rabbinic theme of *da'at Hashem* is essentially a state of mind that the rabbis maintained could be accomplished through the study of Torah. This is why the deliberations and disagreements that occur during the study of Torah are perceived by the rabbis as the pursuit of peace. In the words of the Talmud, "the great scholars of the Torah multiply peace in the world" (Talmud Berachot 64 a).

Perhaps we might illustrate the meaning of this state of mind by turning to two later rabbinic sources. In Ḥasidic texts, the concept of *da'at Hashem* is referred to as *"d'vekut"* which literally means total devotion or 'cleaving' to the divine consciousness. This experience is described in the Ḥasidic masterpiece *The Tanya* in the following way.

> The abode of the divine soul is in the brains that are in the head, and from there it extends to all the limbs . . . It is the source of man's fervent love toward God which, like flaming coals, flares up in the heart of discerning men who understand and reflect with the faculty of knowledge of their brain, on matters that arouse this love; also in the gladness of the heart in the beauty of God and the majesty of His glory which is aroused when the eyes of the wise man, that are in his head i.e. in the brain harboring his wisdom and understanding, gaze at the glory of the King and beauty of His greatness that are unfathomable and without end or limit . . .
>
> In the case of a person who is intelligent enough to know God and to reflect on His blessed greatness and to beget out of his understanding a lofty fear in his brain and a love of God in the right part of his heart so that his soul will thirst for God seeking to cleave unto Him through the fulfillment of the Torah and its commandments.
>
> One binds the mind with a very firm and strong bond to, and firmly fixes his thought on, the greatness of the blessed En Sof (eternity of God) without diverting his mind. For even one who is wise and understanding of the greatness of the blessed En Sof, will not – unless he binds his knowledge and fixes his thought and perseverance – produce in his soul true love and fear, but only vain fancies. Therefore *da'at* is the basis of the virtues and the source of their vitality.[80]

---

**80** Rabbi Shneur Zalman of Liadi, *Likutei Amarim Tanya*, 1797; Bilingual edition, New York: Kehot Publishing, 1973; chapters 3, 9 and 38.

Another version of this same idea appears in the writings of the great modern religious Zionist thinker, Rabbi Abraham Isaac Kook who wrote as follows,

> A person must liberate himself from confinement within his private concerns. This pervades his whole being so that all his thoughts focus only on his own destiny. It reduces him to the worst kind of smallness, and brings upon him endless physical and spiritual distress. It is necessary to raise a person's thought and will and his basic preoccupations toward universality, to the inclusion of all, on the whole world, to man, to the Jewish people, to all existence. This will result in establishing even his private self on a proper basis.
>
> The firmer a person's vision of universality (generality), the greater the joy he will experience and the more he will merit the grace of divine illumination. The reality of God's providence is seen when the world is seen in its totality. For the divine presence cannot rest in a place that is lacking or deficient, a place where there is only a weak point that is constricted and built on nothingness which is individuality and selfness alone.
>
> The way one studies (Torah), the way one prays and the way one shows concern for the rest of the world whether one is concerned with matters that are essential to the narrow existence of man and creations or whether one is concerned with matters that may expand his knowledge (*da'at*) and improve his life, everything assembles and is held together in one unity to complete the general healing of the entire world and to bring it closer and closer to the sought after ideal, to make life good and fair and gradually more and more prepared to connect eternally and be bonded with holiness in a bond of eternal happiness; in a bond of gentle pleasure, bonded to the supernal light, the light of eternal aliveness. He knows what is in the darkness and light dwells with him.[81]

In both the *Tanya* and Rav Kook, the condition of *da'at Hashem* is something to be achieved through the practice and study of Torah. But this practice and study must be accompanied by the right intentionality which perceives of the Torah and its purpose as the expansion of concern about the world from the selfishness of the ego to the totality of all creation – from the separateness of the self to the peaceful unity of all. In Rav Kook's thought, this higher condition, though it is attained through strenuous effort and devotion is also the most natural state of mankind. It is the inner reality that was lost when the fruit of the tree blocked Adam's awareness of the total aliveness and unity of all being and confined it to the narrow and anxious concerns of the self. In Rav Kook's words,

> To be attached to God is the most natural aspiration of a person. What is throughout all existence in a state of dumbness and deafness, in a form of potentiality, is developed in man in a conceptual and experiential form. There can be no substitute in existence for the longing to be absolutely linked with the living God, with the infinite light. As we are under a compulsion to live, to be nourished to grow, so are we under a compulsion to cleave to God.

**81** See Rabbi Kook, *Lights of Holiness*, vol. 3, 199–201 (my translation).

## 5 The Centrality of Prophetic Peace to the Jewish Understanding of Peace

When we consider the centrality of “anti-politics”, “the unity of opposites” and the “knowledge of God” to Jewish thought it feels reasonable to suggest that peace is in fact far more than a key value in Judaism. It is in fact a central organizing principle in ways that this formulation brings to the fore. As we draw towards our conclusion, the point that I wish to make is perhaps an inversion of this i.e. that when we look at Jewish sources that talk explicitly about the concept or value of peace, a huge proportion of them are in fact recognizably referring to the concept of prophetic peace. In other words, the kind of peace that is central to Judaism’s understanding of peace is prophetic rather than legal or political peace.

More specifically, a review of rabbinic sources dealing with peace underlines not only the importance of peace as a value but what might be seen as the rabbis attributing an overwhelmingly and seemingly hyperbolic centrality to peace in the rabbinic understanding of God, the Torah and the world. Here are just a few examples[82]:

> Should you say, “There is food and there is drink but without peace there is nothing!” It is therefore written “And I will put peace in the land” – this teaches us that peace is equivalent to everything. As it is written (Isa. 45:7) “I make light and create darkness, make peace and create everything – this teaches us that peace is equivalent to everything.”[83]

Here the idea that everything is peace connects the very idea of peace with the foundational principles of Torah and creation. Similarly,

> See how great is the power of peace that the holy one blessed be He does not announce the redemption of Jerusalem except in peace as it says, Announce peace (Isa. 52). Alternatively Rabbi Levi said, “Peace is so precious that all the blessings close with peace. The reading of the Shema closes with peace, He who spreads a Succah prayer ends with peace, the priestly blessing closes with peace . . . .Rabbi Shimon Ben Chalafta said Peace is so precious that when the Holy One, blessed be He, asked to bless Israel there was no vessel he could find, which could hold this blessing other than peace. From where do we learn this? As it says, “God will give strength to his people; God will bless his people with peace.[84]

**82** The following excerpts have been translated by me based on a variety of existing English editions.

**83** Sifra Bechukotai 8.

**84** Devarim Rabbah 5,16.

The hyperbolic tone is evident in this passage. Everything in Judaism from the prayers to the redemption of Jerusalem is described as culminating in peace. This text is only an excerpt from a much longer list that recounts the extreme worth of peace in many additional ways. In this context it is important to point out that the final statement is also a culmination in itself because it is quoted by the rabbis who chose the word *shalom* as the last word of the Talmud itself. One final example in this short but illustrative list establishes *shalom* as the foundation of the entire world,

> Avnimos the Guardian asked Rabban Gamliel, "What is the honor (foundation) of the world?" He answered, "Peace". He said to him, "Where is this learned from?" He answered, for it says: "he forms light and creates of darkness makes peace and creates everything. After God created peace he then returned to create everything else. As it says, "steer away from evil and do good seek peace . . ."[85]

Texts that we have not cited here in detail describe *shalom* as a name of God as the name of the Messiah as the name of the people of Israel; as the unity of life and death; as the defining characteristic of the righteous as the completion of the inner soul as the fabric that holds together marriage and the Jewish home, the fabric of the people of Israel and ultimately of the entire world.

Finally, as we draw closer to the implications of prophetic peace to the contemporary reality in the Middle East, it is important to point out how this dual centrality (of peace to Judaism and of prophetic peace to Jewish understandings of peace) is expressed in the writings of Rabbi Abraham Kook.

Notwithstanding his education in the great Lithuanian Talmudic academies,[86] Rabbi Kook's teachings are rooted in Kabbalistic doctrine. His thinking grows out of a concept of the world which contemplates an immanent divine presence in all areas of existence and infers from that universally applicable laws of conduct.[87] It follows, in his view, that the affinities and differences between Israel and the nations of the world are not merely a matter of consciousness and culture;[88] they are

**85** Ps. 34.

**86** Rabbi Naftali Zvi Yehudah Berlin (1816–1893; known by the acronym *Netziv*), the head of the Volozhin Yeshiva, was an important teacher of Rabbi Kook. See Rosenak, *Rabbi Kook*, 11–19 (Hebrew).

**87** For discussion of this approach in contrast to normative sociological thinking, see Rosenak, Avinoam "*Halakhah*. Thought, and the Idea of Holiness in the Writings of Rabbi Haim David Halevi," in: Rachel Elior/Peter Schäfer (eds), *Creation and Re-Creation in Jewish Thought. Festschrift in Honor of Joseph Dan on the Occasion of his Seventieth Birthday*, 309–38, Tübingen: Mohr Siebeck, 2005.

**88** This is the approach found in normative sociological thought; Maimonides was its primary exponent in the Middle Ages.

substantive and ontological.[89] Existence, in all its contradictions, is suffused with the divine presence[90] and those contradictions do not disturb the all-encompassing divine logic.[91] The divine presence instills vitality in the range of spiritual movements and historical processes. This dialectical logic forms the structure for "the doctrine of the unity of opposites" at the center of Rabbi Kook's thinking,[92] a doctrine which as we have seen channels certain key biblical and rabbinic structures through the ideas of the Maharal[93] and through Kabbalistic and Ḥasidic literature overall.[94]

For Rabbi Kook, peace is part of an implicit vision that is less about politics than it is about the discovery of God's oneness in the world. By framing prophecies of peace in messianic time, the prophets leave Jewish history with the legacy of anticipating a future that seems almost impossible. Peace is a culmination of an impossible set of combinations that somehow join together in a unity of opposites that lies beyond human perception. In this perception, Rabbi Kook sees conflict (perhaps ironically) as the result of inadequate variety. Peace comes where human judgment is suspended, where variety is unlimited and no finite combination of subjective truths is allowed to stand for the whole truth. At the heart of the rabbinic project, peace is the product of a limitless process of questioning and classifying applied in a timeless commitment to the endless study and interpretation of every aspect of the law. Like the refraction of light into the colors of the rainbow (through which God expressed his covenant of peace with mankind after the flood), bringing the peace of mankind into the light of day requires the

---

**89** See Tishby, Isaiah, *The Wisdom of the Zohar*, vol. 2, Jerusalem: Mosad Bialik, 1961, 3–93 (Hebrew); Rabbi Judah Halevi, *The Kuzari. An Argument for the Faith of Israel*, trans. Judah Halevi Hartwig Hirschfeld, New York: Schocken Books,1964; part I, sections 26–48, 95.

**90** Rabbi Menachem Mendel of Chernobyl, *Me'or Einayim*, Jerusalem: Me'or Einayim Yeshivah, 1975, 13.

**91** "The force of the contradiction is merely an illness that afflicts logic when limited by the special conditions of man's mind and attentiveness. As we assess the situation, we must sense the contradiction and use that sensation to arrive at a resolution. Above it, however, far above it, there is the supernal divine light, whose possibilities are unlimited and subject to no conditions whatever. It tolerates no impediment on account of the contradiction, and for it, there is no need to resolve it." Kook, Abraham, *Olat Re'ayah*, vol. 1, Jerusalem: Mosad HaRav Kook, 1989, 184.

**92** See note On R. Kook's doctrine of the unity of opposites, see Rosenak, *Rabbi Kook*, 34–42; *Prophetic Halakhah*, 44–57.

**93** Maharal, *Gevurot ha-shem*, 35; Neher, *The Teachings of Maharal*; Rosenak, "Unity of Opposites".

**94** See for example, Kaufman, Tsippi, *Know Him in All Your Ways. The Concept of the Divine and Worship through Corporeality in Early Hasidism*, Ramat-Gan: Bar-Ilan University, 2009, 250–395 (Hebrew).

integration of all shades of opinion. Peace is the culmination of *all* options blended and it is thus an endless quest. An inadequate blend of color produces a murky quality of light, or, in Rabbi Kook's own words,

> Some err to think world peace will be built only through one color, one quality of opinions and characteristics. Therefore when they see that as scholars research the wisdom and knowledge of the Torah through their research different opinion and points of view flourish, they think that causes strife, the opposite of peace. This is really not the case. Real peace can come about only through the value of the flourishing of peace.[95]

This tireless and endless quest for the unity of an ultimately impossible infinite variety of legal options is an echo of the following passage from the *Zohar*,

> Conflict is a distancing of peace, and whoever is in conflict about peace is in disagreement with His holy name, because His holy name is called 'Peace' . . . . Come and behold: the world does not exist except through peace. When the Holy One, blessed be He, created the world, it could not endure until He came and made peace dwell upon them. What is it? It is the Sabbath, which is the peace of the upper and the lower grades. And then the world endured. Therefore, whoever creates dissension about peace will be lost from the world. Rabbi Yosi says that it is written "great peace have they who love your Torah" (Pss. 119,165). The Torah is peace, as it is written "and all her paths are peace" (Prov. 3:17). And Korach came to blemish that peace above . . ." (Zohar, Numbers 16)[96]

The *Zohar* here is distinguishing between a debate conducted for the sake of heaven and one that is not. The former is a debate – or even a conflict – that is motivated (perhaps paradoxically) by the desire for cosmological peace. The notion that peace is the opposite of conflict, though simple, is counter-intuitive to the political worldview in which peace is the by-product of converging interests and agreements. In the political sense peace is what remains to be built when conflict is removed. In the *Zohar's* formulation, peace is a theological term. It is the active antidote to conflict. The conditions required for achieving this kind of peace are in many ways quite the opposite of those needed for political agreements. While conflict resolution requires a compromise – the relinquishing of certain demands in the quest for a common ground or shared system of law – peace is the culmination of infinite differences that must be generated and developed, as it were, from below. Peace must therefore transcend the limits of tolerance and pluralism. It cannot be reached without

**95** Kook, Rabbi Abraham I., "Olat ha-Rayah," in Sagi, *The Open Canon*, 119–122.
**96** "Now Korach," in *The Zohar*, vol. 18, the first unabridged English translation with commentary, ed. and compiled by Rabbi Michael Berg, New York: The Kabbalah Center International Inc., 2003, 225–26.

reclaiming what is ugly, unpleasant and counter-intuitive.[97] It demands the sacrifice of ethics and a radical openness to the impossibility of prophetic surprise. It is the quest that motivates endless generations of study, tireless dedication to minute details, limitless explication, deliberation and dissent – all of which proliferate and ferment, filling pages and pages of rabbinic texts compiled over thousands of years and still expanding in our time.

## 6 Conclusion – The Irenic Irony

I hope it is pertinent and appropriate to follow through on what I said in the introduction to this paper about the potential value of exploring the practical implications of prophetic peace in the context of today's Middle East. I will confine my concluding remarks specifically to the implications that my thesis has for Jewish religious involvement in the peace effort.

It is, I think, important to call attention to what I am calling here the "irenic irony". Beyond being a nice play on words, the irenic irony is a reference to the obvious gap between the centrality and importance of peace in all of the monotheistic religious traditions and the overt participation of religion and religious people in activities that generate and perpetuate conflict. Anyone who asks the crucial question of why the Oslo peace negotiations failed must at least include in

97 It is relevant to mention here the halakhic category of "Avera Lishma": The Talmud suggests in a number of places (e.g. Nazir 23a–b and Brachot 61a) that certain transgressions are necessary for preserving the law. These are categorized as transgressions that are performed in "its name" – in the name of God or in the name of the law. The examples discussed in the Talmud include sexual intercourse between King Ahasuerus and Queen Esther and also the incestuous rape of Lot by his daughters. In both cases, the law is not so much the issue as survival. The Talmudic discussion (Nazir 23b) reaches the somewhat conservative conclusion that such transgressions are accepted in the hope that they will lead to more pious behavior. However, later rabbinic texts toy with the idea that certain transgressions are of spiritual and religious importance in their own right since religious expressions sometimes require transcending the legal boundaries of the Halakha. See for example the stories of the three confessors in the twelfth century text *Sefer Ḥasidim*. In these stories, confessors come to a wise man to describe sins that they have performed, claiming that they only transgressed in order to bring themselves close to temptation so as ultimately to overcome it and repent. But, in order to get there, they needed first to sin. See Ben Samuel, Judah, *Sefer Hasidim*, ed. Jehuda Wistinetzki, Frankfurt/Main: M'kize Nirdamim, 1924, sections 52–53 (Hebrew). Though the confessors are censured for their conduct in this story, later Ḥasidic writers, such as Rabbi Tsadok Hakohen of Lublin, maintain paradoxically that the annulment of the Torah is also its foundation since God's will is served when the law is transgressed in His name.

his or her answer a reference to the fact that both the Israeli and Palestinian leadership faced fierce internal opposition from religious actors. Broadly speaking, beyond political debate and protest this religious opposition has basically been held responsible for fueling settlement activity on the Israeli side and terrorism on the side of the Palestinians. These are perhaps not the only reasons that Oslo failed and they are perhaps not motivated solely by religion, but there can be no doubt of their importance and of the central role that religion played in both.

What I am referring to as the irenic irony here though, runs deeper than the gap between what religion believes and the way religious people act. It also refers to the consistent failure of secular and religious liberals to impress upon those who they see as religious radicals the importance of peace and peaceful behavior as their own tradition demands it. How many times have we heard that "this and that" person's idea of Judaism, Islam or Christianity is not the truth about what that religion *really* teaches?!

Looking at this question solely from the Jewish side of the conflict, it seems quite clear that the failure of liberals and moderates to make an irenic impression on those who's Judaism they criticize can be attributed to a different gap. The irenic irony is perhaps not about the gap between peaceful teaching and peaceful acting so much as it is about the gap between political and prophetic peace. The idea that we began with – i.e. that there is plenty of material in the Jewish tradition that can be called upon to find support for the western secular modes of diplomacy and conflict resolution – has blinded our attention to the fact that the idea of prophetic peace is far more central to both Jewish understandings of peace and of Judaism itself. This is particularly the case in the present historical context; which many religious Jews and almost all national religious Jews identify as a time in which biblical prophecy is gradually and painfully being fulfilled.

In such a time it seems urgent to find ways of seeking peace that align themselves with the principle of prophetic peace or at least make equal space for it in the processes that must inevitably include those whose vision of peace subscribes to it. In a nutshell, this is one of the crucial elements of Siach Shalom (Talking Peace)'s vision.

While academic footnotes have perhaps sufficed to illustrate how Avinoam Rosenak's penetrating insights into the Jewish tradition – and especially into the teachings of Rabbi Kook – have shaped my presentation of prophetic peace, now is the time to speak about the seminal contribution of Sharon Leshem Zinger to the thesis of this paper.[98] She is without doubt a thinker and a true

**98** I want to mention here that I await her forthcoming book whose proposed title is *The Well of Peace – A Dynamic Model for Siach Shalom Talking Peace*.

scholar of Jewish texts but a greater accolade is due to her on account of her ability to uncover how the mechanisms of Jewish thought can give us insight not only into the psyche and the soul but into the complex matrices of interactions that take place between people in both political and social environments. If there is such a thing as a Jewish theory of group dynamics designed to foster the experience of prophetic peace, Leshem Zinger is its author. It is perhaps not customary to accredit in an academic paper lessons that have been learned in an environment of practice, but not to do so in this case would comprise a scholarly crime. The idea that prophetic peace is a workable mode of living together is one that she puts into practice in the group dynamic encounters that we facilitate together. These are built upon the construction of an anti-political environment – a circle of people sitting together as it were around a well – in which starkly opposing points of view can be shared from a place of depth that, when experienced, feels like a spiritual revelation somewhat akin to the knowledge of God. For those whose political outlook is secular the welcoming depths of the interaction are deeply meaningful. But, uniquely in the landscape of the Middle East, these encounters invite the most passionately religious people to feel that their voice is a voice of peace that must be heard.

Given the historical stalemate that peace negotiations have been in for so long, it might perhaps be the time to imagine an anti-political prophetic peace process that is radically inclusive of religious and secular voices in Israeli society today and which might well need to be the face of peace processes in the future.

## Bibliography


Ben Samuel, Judah, *Sefer Hasidim*, ed. Jehuda Wistinetzki, Frankfurt/Main: M'kize Nirdamim, 1924 (Hebrew).

Buber Martin, "The Spirit of Israel and the World Today," in: idem (ed.), *Israel and the World. Essays in a Time of Crisis*, 183–96, Syracuse: Syracuse University Press 1984.

Chernobyl, Rabbi Menachem Mendel of, *Me'or Einayim*, Jerusalem: Me'or Einayim Yeshivah, 1975 (Hebrew).

Derrida, Jacques, "Des tours de Babel," in: *Difference in Translation*, ed. and trans. Joseph P. Graham, Ithaca: Cornell University Press, 1987.

Fisch, Menachem, *Rational Rabbis. Science and Talmudic Culture*, Bloomington: Indiana University Press, 1997.

Gopin, Marc, *Between Eden and Armaageddon*, Oxford: Oxford University Press, 2000.

Arthur, Green, *Seek My Face. A Jewish Mystical Theology*, Woodstock: Jewish Lights Publishing, 2012.

Halbertal, Moshe, *People of the Book. Canon, Meaning and Authority*, Cambridge: Harvard University Press, 1997.

Halevi, Rabbi Judah, *The Kuzari. An Argument for the Faith of Israel*, trans. Judah Halevi Hartwig Hirschfeld, New York: Schocken Books, 1964.

Hartman, David, *The Living Covenant*, New York: Jewish Lights Press, 1980.

Heschel, Rabbi Abraham J., *The Sabbath*, New York: Farrar Strauss and Giroux, 1951.

Heschel, Rabbi Abraham J., *The Prophets*, New York: Harper and Row, 1962.

Hirsch, Samson R., *The Pentateuch. With Translation and Commentary*, New York: Judaica Press, 1962. Reissued in a new translation as Haberman, Daniel, *The Hirsch Chumash*, New York: Feldheim/Judaica Press, 2009.

Isaacs, Alick, *A Prophetic Peace. Judaism, Religion and Politics*, Bloomington: Indiana University Press, 2011.

Kaminsky, Howard Gary, *Traditional Jewish Perspectives on Peace and Interpersonal Conflict Resolution*, New York: Teachers College Columbia University, 2005.

Kaplan, Aryeh, *Inner Space*, New York: Moznaim Publishing, 1990.

Kaufman, Tsippi, *Know Him in All Your* Ways. *The Concept of the Divine and Worship through Corporeality in early Hasidism*, Ramat-Gan: Bar-Ilan University, 2009 (Hebrew).

Kook, Rabbi Abraham, *Lights of Holiness*, vol. 3, Jerusalem: Mosad HaRav Kook, 1985 (Hebrew).

Konrad, George, *Anti-Politics. An Essay*, trans. Richard E. Allen, New York: Quartet, 1984.

Leshem-Zinger, Sharon, *The Vessel of Peace. A Cultural and Symbolic Model for Making Decisions Peacefully about Peace* (forthcoming in Hebrew).

Liadi, Rabbi Shneur Zalman of, *Likutei Amarim Tanya*, 1797, bilingual edition, New York: Kehot Publishing, 1973.

Neher, André, *The Teachings of Maharal*, Jerusalem: Reuben Mass, 2003 (Hebrew).

Roness, Michal (ed.), *Conflict and Conflict Management in Jewish Sources*, Ramat Gan: Bar Ilan University Program on Conflict Management and Negotiation, 2008.

Rosenak, Avinoam, "Hidden Diaries and New Discoveries. The Life and Thought of Rabbi A.I. Kook," *Shofar. An Interdisciplinary Journal of Jewish Studies* 25:3 (2007), 111–47.

Rosenak, Avinoam, *Prophetic Halakha. The Philosophy of Halakha in the Teaching of Rav Kook*, Jerusalem: The Magnes Press, 2007 (Hebrew).

Rosenak, Avinoam, "Unity of Opposites in the Teachings of Maharal. A Study of his Writings and Their Implications for Jewish Thought in the Twentieth and Twenty-first Centuries," in: Elchanan Reiner (ed.), *Akdamot*, 449–87, Jerusalem: Zalman Shazar Institute, 2015 (Hebrew).

Rosenak, Avinoam, "*Halakhah*. Thought, and the Idea of Holiness in the Writings of Rabbi Haim David Halevi," in: Rachel Elior/Peter Schäfer (eds), *Creation and Re-Creation in Jewish* Thought. *Festschrift in Honor of Joseph Dan on the Occasion of his Seventieth Birthday*, 309–38, Tübingen: Mohr Siebeck, 2005.

Rosenak, Avinoam, *Rabbi Kook*, Jerusalem: Zalman Shazar Center, 2006 (Hebrew).

Roth, Daniel, *The Tradition of Aaron Pursuer of Peace Between People as a Rabbinic Model of Reconciliation*, PhD diss. Bar Ilan University 2012.

Sagi, Avi (ed.), *The Open Canon. On the Meaning of Halakhic Discourse*, trans. Batya Stein, London/New York: Continuum, 2007.

Soloveitchik, Joseph Dov (1903–1993), *Kol Dodi Dofek* (*Fate and Destiny. From the Holocaust to the State of Israel*), New York: Ktav Publishing House, 2000.

Sommer, Benjamin, "Revelation at Sinai in the Hebrew Bible and in Jewish Theology," *The Journal of Religion* 79:3 (1999), 422–51.

Steinberg, Gerald M., "Jewish Sources on Conflict Management. Realism and Human Nature," in: Roness, Michael (ed.), *Conflict and Conflict Management in Jewish Sources*, Ramat Gan: Bar Ilan University Program on Conflict Management and Negotiation, 2008, 10–23.
Sutton, Avraham, *Spiritual Technology*, New York: Hebrewbooks, 2013.
Tishby, Isaiah, *The Wisdom of the Zohar*, vol. 2, Jerusalem: Mosad Bialik, 1961 (Hebrew).
Weinreb, Friedrich, *Roots of the Bible. An Ancient View for a New Outlook*, trans. N. Keus, Braunton: Melrin Books, 1986.
*The Zohar*, vol. 18, the first unabridged English translation with commentary, ed. and compiled by Rabbi Michael Berg, New York: The Kabbalah Center International Inc., 2003.

Volker Stümke

# The Concept of Peace in Christianity

Peace – who among us does not desire to live without war and struggles, without need and fear? Considering peace in this comprehensive way, most would admit that we are not living in peaceful times. Furthermore, these desires were described in the negative ("without") – how to name the opposite, the affirmative features of peace, is indeed debated. As in most religions the Christian faith has developed an understanding of peace. For Christians this is grounded in the Bible as Holy Scripture and can be extended through history of the Church, and is now facing today's challenges. One of the basic assertions concerning peace is found in the Hebrew Bible:

> I will hear what God the LORD will speak: for he will speak peace to his people, and to his saints: but let them not turn again to folly. Surely his salvation is near them that fear him; that glory may dwell in our land. Mercy and truth are met together; righteousness and peace have kissed each other. Truth shall spring out of the earth; and righteousness shall look down from heaven. Yes, the LORD shall give that which is good; and our land shall yield her increase. Righteousness shall go before him; and shall set us in the way of his steps.
>
> (Ps. 85:8–13)[1]

Accordingly, in Christian understanding, peace, the righteousness of the people and justice in the land must be inseparably connected. Peace is not only a personal experience, but a political and social achievement as well. To unfold the Christian understanding of peace, this chapter is divided in three sections, each containing four subsections. The sections follow the historical development starting [1.] with the Bible, then [2.] going through some main insights in Church history, and ending [3.] with the current tasks. The four subsections address the different aspects of peace; namely: the political, the social, the personal, and the religious.

## 1 Biblical References Regarding Peace

For Christians, the Hebrew Bible (also called the Old Testament) and the New Testament together form the Holy Scriptures. This collection of writings came into being in a process that took about eight hundred years, so there are a lot of

1 The quotations from the Bible follow The American King James Version.

https://doi.org/10.1515/9783110682021-002

descriptions of peace and there is a development of insights concerning living together without war, violence, need, and fear. Firstly, peace is interpreted in a positive and in a negative way; so the prevalent distinction between positive and negative peace in the political sciences is also found in the Bible.[2] Peace (שלום / shalom) is described in the Hebrew Bible negatively as the absence of war, and positively as living together as God's chosen people contently and safely, protected against defamation and false accusations (Ps. 4).[3] Perhaps God's salvation experienced on earth is the best paraphrase for peace. The New Testament mostly adopted this understanding of peace (*εἰρήνη* / eirene). Peace is as well negatively understood as the absence of war, and positively unfolded as the reconciled relationship between God and humans through Jesus Christ and out of this as the virtue of brotherly love and humility.[4]

Secondly, the Biblical authors compare these desired conditions with their experience in the real world. In the real world this positive peace is not naturally granted. Instead, the Biblical scriptures stress the fact that violence has been ruling this world from the beginning, when Cain slew his brother Abel (Gen. 4). As people often harm each other, even negative peace is rarely found.[5]

---

**2** Cf. for the following Otto, Eckhart, *Krieg und Frieden in der Hebräischen Bibel und im Alten Orient. Aspekte für eine Friedensordnung in der Moderne*, Stuttgart: Kohlhammer, 1999; Krochmalnik, Daniel, "Krieg und Frieden in der hebräischen Bibel und rabbinischen Traditionen," in: Ines-Jacqueline Werkner/Klaus Ebeling (eds), *Handbuch Friedensethik*, 191–202, Wiesbaden: Springer, 2017; Schnocks, Johannes, *Das Alte Testament und die Gewalt. Studien zu göttlicher und menschlicher Gewalt in alttestamentlichen Texten und ihren Rezeptionen*, Neukirchen-Vluyn: Neukirchener Verlagsgesellschaft, 2014; Schwienhorst-Schönberger, Ludger, "Recht und Gewalt im Alten Testament," in: Nadja Rossmanith et al. (eds), *Sprachen heiliger Schriften und ihre Auslegung*, 7–33, Institut für Religion und Frieden (Ethica Themen), Wien: BMLVS Heeresdruckerei, 2015.

**3** **Ps. 4**: "Hear me when I call, O God of my righteousness: you have enlarged me when I was in distress; have mercy on me, and hear my prayer. O you sons of men, how long will you turn my glory into shame? How long will you love vanity, and seek after leasing? [. . .] But know that the LORD has set apart him that is godly for himself: the LORD will hear when I call to him. Stand in awe, and sin not: commune with your own heart on your bed, and be still. [. . .] Offer the sacrifices of righteousness, and put your trust in the LORD. There be many that say: Who will show us any good? LORD, lift you up the light of your countenance on us. You have put gladness in my heart, more than in the time that their corn and their wine increased. I will both lay me down in peace, and sleep: for you, LORD, only make me dwell in safety."

**4** Cf. Forderer, Tanja, "Frieden in den neutestamentlichen Schriften," in: Elisabeth Gräb-Schmidt/Julian Zeyher-Quattlender (eds), *Friedensethik und Theologie. Systematische Erschließung eines Fachgebiets aus der Perspektive von Philosophie und christlicher Theologie*, 117–36, Baden-Baden: Nomos, 2018.

**5** Cf. Die deutschen Bischöfe, *Gerechter Friede*, Bonn: Sekretariat der Deutschen Bischofskonferenz, 2000, first chapter; Baumann, Gerlinde, "Gewalt in biblischen Texten.

Israel experienced many wars during its history. In the year 587 BC, the capital Jerusalem, the temple, and the entire state of Israel were destroyed, and the people were sent into the Babylonian exile. Thus the desire for peace and the experience of violence contradict each other. Nevertheless, there is a vivid hope in the Scriptures not only for negative, but also for positive peace. This positive peace is characterized by just conditions for everyone and not only by the absence of war and violence. Living together in peace indicates that neither poverty nor exploitation nor breaching of contracts will occur (Isa. 11:3–5), but that there will be harmony among peoples and that they will live in accordance with nature (Isa. 11:6–9). The Lord of justice will eliminate the wrongdoers, so that tranquillity and security will reign (Isa. 32:15–18).[6] And God will overcome poverty and need (Ps. 9:16–18).[7]

Thirdly, the main point in awaiting this desired peace is its dependence on God. As mankind is weak and sinful, we are not able to put these paradisiac conditions into execution. It is the Lord who will bring peace. From the initial fratricide, God has helped humans in limiting violence. He marked Cain allowing no one to kill him (Gen. 4:15); by this God prevents a cycle of violence. Later, God chose Abraham and then Moses as his partners, making a covenant with them representing his chosen people. The rules in this covenant were also limiting the use of force. For example, the prescription "an eye for an eye and a tooth for a tooth" (Exod. 21:23–25) hinders an exaggeration of violence in prosecution by stressing the proportionality of harmful answers. These limitations helped establishing and keeping the negative peace, although they cannot bring positive peace – this will be achieved by God's Messiah, a chosen messenger of God that will fulfill the promise of living together contently and safely:

> And there shall come forth a rod out of the stem of Jesse, and a Branch shall grow out of his roots: And the spirit of the LORD shall rest on him, the spirit of wisdom and understanding, the spirit of counsel and might, the spirit of knowledge and of the fear of the

---

Hintergründe, Differenzierungen, hermeneutische Überlegungen," in: Severin J. Lederhilger (ed.), *Gewalt im Namen Gottes. Die Verantwortung der Religionen für Krieg und Frieden*, 83–95, Frankfurt/Main: Peter Lang, 2015.

6 **Isa. 32:15–18**: "Until the spirit be poured on us from on high, and the wilderness be a fruitful field, and the fruitful field be counted for a forest. Then judgment shall dwell in the wilderness, and righteousness remain in the fruitful field. And the work of righteousness shall be peace; and the effect of righteousness quietness and assurance for ever. And my people shall dwell in a peaceable habitation, and in sure dwellings, and in quiet resting places."

7 **Ps. 9:16–18**: "The LORD is known by the judgment which he executes: the wicked is snared in the work of his own hands. [. . .] The wicked shall be turned into hell, and all the nations that forget God. For the needy shall not always be forgotten: the expectation of the poor shall not perish for ever."

> LORD; and shall make him of quick understanding in the fear of the LORD: and he shall not judge after the sight of his eyes, neither reprove after the hearing of his ears: But with righteousness shall he judge the poor, and reprove with equity for the meek of the earth: and he shall smite the earth: with the rod of his mouth, and with the breath of his lips shall he slay the wicked. And righteousness shall be the girdle of his loins, and faithfulness the girdle of his reins. The wolf also shall dwell with the lamb, and the leopard shall lie down with the kid; and the calf and the young lion and the fatted calf together; and a little child shall lead them. And the cow and the bear shall feed; their young ones shall lie down together: and the lion shall eat straw like the ox. And the sucking child shall play on the hole of the asp, and the weaned child shall put his hand on the cockatrice' den. They shall not hurt nor destroy in all my holy mountain: for the earth shall be full of the knowledge of the LORD, as the waters cover the sea. (Isa. 11:1–9)

This profound peace is a vivid hope for Christians (and for Jews as well) and it will not be realized by humans but by God via his Messiah. He, the Prince of Peace, will not only bring violence to an end (negative peace) but will furthermore create this ideal world in God's authority (Isa. 9:1–5).[8] Thus, positive peace is a hope for the future. God will certainly intervene and will reward the faithful (Deut. 12:1–12).[9]

---

8 **Isa. 9:1–5**: "Nevertheless the dimness shall not be such as was in her vexation, when at the first he lightly afflicted the land of Zebulun and the land of Naphtali, and afterward did more grievously afflict her by the way of the sea, beyond Jordan, in Galilee of the nations. The people that walked in darkness have seen a great light: they that dwell in the land of the shadow of death, on them has the light shined. You have multiplied the nation, and not increased the joy: they joy before you according to the joy in harvest, and as men rejoice when they divide the spoil. For you have broken the yoke of his burden, and the staff of his shoulder, the rod of his oppressor, as in the day of Midian. For every battle of the warrior is with confused noise, and garments rolled in blood; but this shall be with burning and fuel of fire."

9 **Deut. 12:1–12**: And at that time shall Michael stand up, the great prince which stands for the children of your people: and there shall be a time of trouble, such as never was since there was a nation even to that same time: and at that time your people shall be delivered, every one that shall be found written in the book. And many of them that sleep in the dust of the earth shall awake, some to everlasting life, and some to shame and everlasting contempt. And they that be wise shall shine as the brightness of the firmament; and they that turn many to righteousness as the stars for ever and ever. But you, O Daniel, shut up the words, and seal the book, even to the time of the end: many shall run to and fro, and knowledge shall be increased. Then I Daniel looked, and, behold, there stood other two, the one on this side of the bank of the river, and the other on that side of the bank of the river. And one said to the man clothed in linen, which was on the waters of the river: How long shall it be to the end of these wonders? And I heard the man clothed in linen, which was on the waters of the river, when he held up his right hand and his left hand to heaven, and swore by him that lives for ever that it shall be for a time, times, and an half; and when he shall have accomplished to scatter the power of the holy people, all these things shall be finished. And I heard, but I understood not:

Then God's enemies will be gone completely and swords will be hammered to ploughshares (Mic. 4:1–4).[10]

According to Col. 1:15–20[11] and Luke 4:16–21,[12] Christians actually believe that Jesus is this Messiah ("Christ" is the Greek translation of the Hebrew term "Messiah"), whereas Jews do not agree, and therefore still wait for the arrival of the redeemer. At this juncture, the main difference between the Jewish and the Christian understanding of peace arises. As Christians are convinced that Jesus is the Christ, they also claim that the negative and the promised positive peace are already accessible here on earth. To emphasize this, Christians refer mainly to the

---

then said I, O my Lord, what shall be the end of these things? And he said, Go your way, Daniel: for the words are closed up and sealed till the time of the end. Many shall be purified, and made white, and tried; but the wicked shall do wickedly: and none of the wicked shall understand; but the wise shall understand. And from the time that the daily sacrifice shall be taken away, and the abomination that makes desolate set up, there shall be a thousand two hundred and ninety days. Blessed is he that waits, and comes to the thousand three hundred and five and thirty days.

**10 Mic. 4:1–4**: But in the last days it shall come to pass, that the mountain of the house of the LORD shall be established in the top of the mountains, and it shall be exalted above the hills; and people shall flow to it. And many nations shall come, and say, Come, and let us go up to the mountain of the LORD, and to the house of the God of Jacob; and he will teach us of his ways, and we will walk in his paths: for the law shall go forth of Zion, and the word of the LORD from Jerusalem. And he shall judge among many people, and rebuke strong nations afar off; and they shall beat their swords into plowshares, and their spears into pruning hooks: nation shall not lift up a sword against nation, neither shall they learn war any more. But they shall sit every man under his vine and under his fig tree; and none shall make them afraid: for the mouth of the LORD of hosts has spoken it.

**11 Col. 1:15–20**: [Christ,] "Who is the image of the invisible God, the firstborn of every creature: For by him were all things created, that are in heaven, and that are in earth, visible and invisible, whether they be thrones, or dominions, or principalities, or powers: all things were created by him, and for him: And he is before all things, and by him all things consist. And he is the head of the body, the church: who is the beginning, the firstborn from the dead; that in all things he might have the preeminence. For it pleased the Father that in him should all fullness dwell; And, having made peace through the blood of his cross, by him to reconcile all things to himself; by him, I say, whether they be things in earth, or things in heaven."

**12 Luke 4:16–21**: "And he came to Nazareth, where he had been brought up: and, as his custom was, he went into the synagogue on the sabbath day, and stood up for to read. And there was delivered to him the book of the prophet Esaias. And when he had opened the book, he found the place where it was written, the Spirit of the Lord is on me, because he has anointed me to preach the gospel to the poor; he has sent me to heal the brokenhearted, to preach deliverance to the captives, and recovering of sight to the blind, to set at liberty them that are bruised, to preach the acceptable year of the Lord. And he closed the book, and he gave it again to the minister, and sat down. And the eyes of all them that were in the synagogue were fastened on him. And he began to say to them, This day is this scripture fulfilled in your ears."

religious, the personal and the social perspective, whereas Jews stress the fact that in the political perspective peace has not yet arrived on earth. This implies that the Messiah did not yet appear. I will now focus on the Christian interpretation.

## 1.1 From the Religious Perspective

In his letter to the Romans, apostle Paul outlines the salvation Jesus brought to humans in reference to the peace that is established through Jesus as the Christ:

> Therefore being justified by faith, we have peace with God through our Lord Jesus Christ: By whom also we have access by faith into this grace wherein we stand, and rejoice in hope of the glory of God. And not only so, but we glory in tribulations also: knowing that tribulation works patience; and patience, experience; and experience, hope: And hope makes not ashamed; because the love of God is shed abroad in our hearts by the Holy Ghost which is given to us. (Rom. 5:1–5)

According to Paul, peace with God is the main aim that Jesus has achieved. In Jesus Christ God overcomes violence with love and reconciliation. Whilst violence can merely be limited by other violence, and even this only as long as it is used proportionally, God's love is able to bring violence to an end and to establish the positive peace. Since this love is stronger than the sin of humans, love can drain the sources of violence. According to most of the authors of the New Testament, including Paul, sin as a broken relationship of all humans to God, and indeed causes evil deeds such as violence (Rom. 3:11–18).[13] Furthermore, they are convinced that humans are too weak to overcome these sins by themselves; God himself has rendered redemption through Jesus Christ instead (Rom. 3:23–24). The itinerant preacher Jesus from Nazareth has preached God to be a merciful father, willing to forgive all sins (Luke 15:11–24). Thus, only repentance and faith are needed to be redeemed. After being condemned and crucified as a criminal by the political and religious leaders, Jesus was resurrected by God. Ipso facto it was evident for the believers that those leaders were wrong, whereas Jesus was in the right and had proclaimed God as he really is (Acts 2:23f). This implies

13 **Rom. 3:11–18**: "There is none that understands, there is none that seeks after God. They are all gone out of the way, they are together become unprofitable; there is none that does good, no, not one. Their throat is an open sepulchre; with their tongues they have used deceit; the poison of asps is under their lips: Whose mouth is full of cursing and bitterness: Their feet are swift to shed blood: Destruction and misery are in their ways: And the way of peace have they not known: There is no fear of God before their eyes."

that the forgiveness of sins has become reality and that the hostility between God and the sinners has been overcome (1John 4:8–11).[14]

As a result, positive peace within Christian communities can flourish on the foundation that was laid by Jesus Christ (1Cor. 3:5–11).[15] Jesus has broken down the walls not only between God and humans, but those between humans as well: they are also reconciled with each other and by this can live together contently and safely (Eph. 2:13–19).[16] Consequently, in the Christian communities the religious controversies between humans shall come to an end, since they all have united in Jesus the Christ as the one body with many different limbs (1Cor. 12:12–14),[17] although their unification neither implies giving up all individual characteristics nor denying everyone's specific spiritual gifts (1Cor.12:4–7).

This is, of course, an idealistic description that neither in the Ancient Church nor through history has become reality. Most of Paul's letters are dealing with controversies already in the first communities. These debates were partly about theological quarrels, how to understand the creed "Jesus is the Christ" and to unfold the implications, but they were also approaching concrete problems about the living together in a Christian community and in a non-Christian surrounding. The presence of such controversies was interpreted as evidence that sin is still

---

14 **1John 4:8–11**: "He that loves not knows not God; for God is love. In this was manifested the love of God toward us, because that God sent his only begotten Son into the world, that we might live through him. Herein is love, not that we loved God, but that he loved us, and sent his Son to be the propitiation for our sins. Beloved, if God so loved us, we ought also to love one another."

15 **1Cor. 3:5–11**: "Who then is Paul, and who is Apollos, but ministers through whom you believed, as the Lord gave to each one? I planted, Apollos watered, but God gave the increase. So then neither he who plants is anything, nor he who waters, but God who gives the increase. Now he who plants and he who waters are one, and each one will receive his own reward according to his own labour. For we are God's fellow workers; you are God's field, you are God's building. According to the grace of God which was given to me, as a wise master builder I have laid the foundation, and another builds on it. But let each one take heed how he builds on it. For no other foundation can anyone lay than that which is laid, which is Jesus Christ."

16 **Eph. 2:13–19**: "But now in Christ Jesus you who sometimes were far off are made near by the blood of Christ. For he is our peace, who has made both one, and has broken down the middle wall of partition between us; Having abolished in his flesh the enmity, even the law of commandments contained in ordinances; for to make in himself of two one new man, so making peace; And that he might reconcile both to God in one body by the cross, having slain the enmity thereby: And came and preached peace to you which were afar off, and to them that were near. For through him we both have access by one Spirit to the Father. Now therefore you are no more strangers and foreigners, but fellow citizens with the saints, and of the household of God."

17 **1Cor. 12:12–14**: "For as the body is one, and has many members, and all the members of that one body, being many, are one body: so also is Christ. For by one Spirit are we all baptized into one body, whether we be Jews or Gentiles, whether we be bond or free; and have been all made to drink into one Spirit. For the body is not one member, but many."

there and still mighty, even amongst the Christians, who should not deny this (1John 1:8–9).[18] The realization of peace is a process, having started with the resurrection of Jesus Christ and now leading to the Kingdom of God. Not only humans, but the whole of creation is eagerly waiting for this salvation (Rom. 8:19–23).[19]

Furthermore, there are still quarrels between religions and there are still many who do not believe in Jesus as the Christ, which the New Testament clearly admits. Thus Paul calls on all humans to become reconciled with God (2Cor. 5:17–21).[20] Reconciliation and Salvation are, so to speak, in progress; the Son of God and the Holy Spirit testify these deeds of God and use the apostles and other Christians as their fellow workers (1Cor. 3:9; 1Thess 3:2). When the work is fulfilled, God will rule and will be all in all (1Cor. 15:28). So it is the future hope for Christians that peace with God as the starting point will spread all over the world – then the Kingdom of God will be realized and the positive peace will become universal reality.

## 1.2 From the Personal Perspective

Being reconciled and having found peace with God has impacts for the Christians. The main effect of trusting in Jesus Christ is to be content with one's own life. Christians are satisfied and joyful, because they live in certainty of faith: nothing will be able to separate them from the love of God (Rom. 8:31–39).[21] They have all they need to gain everlasting peace. The certainty of faith that God is the merciful

18 **1John 1:8–9**: "If we say that we have no sin, we deceive ourselves, and the truth is not in us. If we confess our sins, he is faithful and just to forgive us our sins, and to cleanse us from all unrighteousness."

19 **Rom. 8:19–23**: "For the earnest expectation of the creature waits for the manifestation of the sons of God. For the creature was made subject to vanity, not willingly, but by reason of him who has subjected the same in hope, because the creature itself also shall be delivered from the bondage of corruption into the glorious liberty of the children of God. For we know that the whole creation groans and travails in pain together until now. And not only they, but ourselves also, which have the first fruits of the Spirit, even we ourselves groan within ourselves, waiting for the adoption, to wit, the redemption of our body."

20 **2Cor. 5:17–21**: "Therefore if any man be in Christ, he is a new creature: old things are passed away; behold, all things are become new. And all things are of God, who has reconciled us to himself by Jesus Christ, and has given to us the ministry of reconciliation; To wit, that God was in Christ, reconciling the world to himself, not imputing their trespasses to them; and has committed to us the word of reconciliation. Now then we are ambassadors for Christ, as though God did beseech you by us: we pray you in Christ's stead, be you reconciled to God. For he has made him to be sin for us, who knew no sin; that we might be made the righteousness of God in him."

21 **Rom. 8:31–39**: "What shall we then say to these things? If God be for us, who can be against us? He that spared not his own Son, but delivered him up for us all, how shall he not with him

father is sufficient to lead a content and safe life on earth (John 14:8).[22] No efforts before God are needed, for example, neither to circumcise nor to observe dietary laws is necessary to be accepted by God. It has become superfluous to distinguish oneself in front of God and at the expense of other humans; the way to heaven is not a competition. Instead, Christians are liberated from such a self-centered perspective and thus became able to focus on neighborly love – and shall henceforth do this. In his letter to the Galatians, Paul expressed this insight as follows:

> Stand fast therefore in the liberty with which Christ has made us free, and be not entangled again with the yoke of bondage. Behold, I Paul say to you, that if you be circumcised, Christ shall profit you nothing. For I testify again to every man that is circumcised, that he is a debtor to do the whole law. Christ is become of no effect to you, whoever of you are justified by the law; you are fallen from grace. For we through the Spirit wait for the hope of righteousness by faith. For in Jesus Christ neither circumcision avails anything, nor uncircumcision; but faith which works by love. You did run well; who did hinder you that you should not obey the truth? This persuasion comes not of him that calls you. A little leaven leavens the whole lump. I have confidence in you through the Lord, that you will be none otherwise minded: but he that troubles you shall bear his judgment, whoever he be. And I, brothers, if I yet preach circumcision, why do I yet suffer persecution? Then is the offense of the cross ceased. I would they were even cut off which trouble you. For, brothers, you have been called to liberty; only use not liberty for an occasion to the flesh, but by love serve one another. For all the law is fulfilled in one word, even in this; You shall love your neighbour as yourself. (Gal. 5:1–14)

The twofold law of love (Matt. 22:35–40)[23] is the principle rule for Christians to obey. They know on the one hand that they have come into this world with nothing and that they will carry nothing out (1Tim. 6:6–7).[24] On the other hand

---

also freely give us all things? Who shall lay any thing to the charge of God's elect? It is God that justifies. Who is he that condemns? It is Christ that died, yes rather, that is risen again, who is even at the right hand of God, who also makes intercession for us. Who shall separate us from the love of Christ? shall tribulation, or distress, or persecution, or famine, or nakedness, or peril, or sword? As it is written, For your sake we are killed all the day long; we are accounted as sheep for the slaughter. No, in all these things we are more than conquerors through him that loved us. For I am persuaded, that neither death, nor life, nor angels, nor principalities, nor powers, nor things present, nor things to come, Nor height, nor depth, nor any other creature, shall be able to separate us from the love of God, which is in Christ Jesus our Lord."

**22 John 14:8**: "Philip said to him, Lord, show us the Father, and it suffices us."

**23 Matt. 22:35–40**: "Then one of them [Pharisees], which was a lawyer, asked him a question, tempting him, and saying, Master, which is the great commandment in the law? Jesus said to him: You shall love the Lord your God with all your heart, and with all your soul, and with all your mind. This is the first and great commandment. And the second is like to it: You shall love your neighbor as yourself. On these two commandments hang all the law and the prophets."

**24 1Tim. 6:6–8**: "But godliness with contentment is great gain. For we brought nothing into this world, and it is certain we can carry nothing out. And having food and raiment let us be therewith content."

they can accept these conditions in humility and will engage for the benefit of their neighbors (Rom. 12:16–21),[25] since this "nothing" is enough facing the merciful God.

The engagement of Christians shall support peace. This implies mainly to remain in the freedom of faith. Freed through Jesus Christ they tend not to be overruled again by other powers. On the one hand, this denotes self-discipline to avoid a fall back, on the other hand it implies living thankfully and peacefully in the parishes (Col. 3:12–16).[26] Christians shall foster justness by neither oppressing nor exploiting nor betraying their neighbors. Up to now, controlling one's emotions and engaging in humanities are central means for a society to live together peacefully, as Dieter Senghaas has analyzed.[27] Enviousness and hate are expendable for Christians, since injustice and inequity cannot set aside God's promise. Christians do not deny that injustice, poverty, misery and violence still exist. However, facing this, Christians are not obsessed with their own well-being and salvation; they are exempt from these selfish concerns and therefore can engage for the welfare of their neighbors. Ideally, Christians will be content with what they have and will engage in favor of others who suffer under bad conditions. Furthermore, they will not lament the trouble they are undergoing, but will stand up for diminishing the pain of their neighbor. By this means, reconciliation and peace become reality – fragmentally and symbolically. And the Christians following Jesus Christ act towards their neighbor just like Christ acted towards them.

---

**25 Rom. 12:16–21**: "Be of the same mind one toward another. Mind not high things, but condescend to men of low estate. Be not wise in your own conceits. Recompense to no man evil for evil. Provide things honest in the sight of all men. If it be possible, as much as lies in you, live peaceably with all men. Dearly beloved, avenge not yourselves, but rather give place to wrath: for it is written, Vengeance is mine; I will repay, said the Lord. Therefore if your enemy hunger, feed him; if he thirst, give him drink: for in so doing you shall heap coals of fire on his head. Be not overcome of evil, but overcome evil with good."

**26 Col. 3:12–16**: "Put on therefore, as the elect of God, holy and beloved, bowels of mercies, kindness, humbleness of mind, meekness, long-suffering; Forbearing one another, and forgiving one another, if any man have a quarrel against any: even as Christ forgave you, so also do you. And above all these things put on charity, which is the bond of perfection. And let the peace of God rule in your hearts, to the which also you are called in one body; and be you thankful. Let the word of Christ dwell in you richly in all wisdom; teaching and admonishing one another in psalms and hymns and spiritual songs, singing with grace in your hearts to the Lord."

**27** Cf. Senghaas, Dieter, "Frieden als Zivilisierungsprojekt," in: id., *Den Frieden denken. Si vis pacem, para pacem*, 196–223, Frankfurt/Main: Suhrkamp, 1975. Senghaas listed six features in his civilizational hexagon that support social peace, namely: the monopoly on the use of force, the rule of law, democratic participation, social justice (including engagement in humanities), a constructive conflict culture, and control of emotion.

However, in the Christian perspective, these social improvements are grounded in the peaceful mind of the believers. Therefore, the New Testament stresses that Christians are able to and shall seek and ensue peace (1Pet. 3: 10–11).[28] This refers to the blessing Jesus spoke out in his sermon on the mount: "Blessed are the peacemakers: for they shall be called the children of God" (Matt. 5:9). Gifted with the divine peace Christians will support peace on earth by serving the others (Mark 10:42–45).[29]

## 1.3 From the Social Perspective

The peace-engagement of the Christians, grounded in their certitude of faith, is oriented not only to live together in a parish or a community, but to foster the commonwealth of their society. The Hebrew Bible already expects God or God's anointed (the Messiah) to overcome poverty and need (Ps. 72:1–4)[30] so that everyone will own land and will be able to fend for himself (Deut. 15:4).[31] The New Testament picks up this expectation and simultaneously modifies the realization: While Jesus the Christ has already reconciled humans with God, it is now up to Christians as fellow workers of God's Son to help realize this peace in society (Col. 4:11). As already mentioned, the Christians follow the law of love by practicing neighborly love and even love for enemies. And in the congregations, especially in the service they bear the tension between the reconciliation that is already there and the eternal peace that is still outstanding.

This love became reality firstly in the personal relations of the congregations, but was not limited to this. More and more the brotherly love of Christians spread

---

**28 1Pet. 3:10–11**: "For he that will love life, and see good days, let him refrain his tongue from evil, and his lips that they speak no guile: Let him eschew evil, and do good; let him seek peace, and ensue it."

**29 Mark 10:42–45**: "But Jesus called them to him, and said to them, You know that they which are accounted to rule over the Gentiles exercise lordship over them; and their great ones exercise authority on them. But so shall it not be among you: but whoever will be great among you, shall be your minister: And whoever of you will be the most chief, shall be servant of all. For even the Son of man came not to be ministered to, but to minister, and to give his life a ransom for many."

**30 Ps. 72:1–4**: "Give the king your judgments, O God, and your righteousness to the king's son. He shall judge your people with righteousness, and your poor with judgment. The mountains shall bring peace to the people, and the little hills, by righteousness. He shall judge the poor of the people, he shall save the children of the needy, and shall break in pieces the oppressor."

**31 Deut. 15:4**: "Save when there shall be no poor among you; for the LORD shall greatly bless you in the land which the LORD your God gives you for an inheritance to possess it."

out into society. In the times of the Ancient Church, the engagement for social welfare was a specific feature of the Christians. They promoted solidarity by helping the poor, the old, the ill, and the disabled.[32] Furthermore, they ended the isolation from those people who were living at the edge of society, as Jesus had done by healing the ill (Mark 6:53–56) and having meal with the sinners and the tax-collectors (Luke 15:1–2). In the congregations, the rich Christians were exhorted not to become arrogant but charitable (1Tim. 6:17–19).[33] Besides this, Christians should realise soberly that this progression in society may be helpful, but will not suffice to establish extensive peace. After all, they should support in word and deed the peace God has accomplished through his love and reconciliation, since God's deed has overcome the enmity between God and mankind. Christian life and action is therefore meant to be a symbolic re–presentation of this peace.

Thus, righteousness (*δικαιοσύνη* / dikaiosynai) is the most important word the New Testament uses in this context. This term comprises three meanings: Firstly, it means the redemption by God (Rom. 3:21–24).[34] God redeems the sinners through the crucifixion of his Son, so Christians will pass Judgment Day. Secondly, this term denotes the right behavior of Christians, their righteousness that was imputed to them (as already to Abraham) through their faith (Rom. 4:20–25).[35] And thirdly, it includes, in the Hebrew Bible as well, a just

---

**32** Cf. Gerhard K. Schäfer/ Volker Herrmann, Geschichtliche Entwicklungen der Diakonie von der Alten Kirche bis zur Gegenwart im Überblick; in: Volker Herrmann/ Martin Horstmann (eds), Studienbuch Diakonik Band 1: biblische, historische und theologische Zugänge zur Diakonie, Neukirchen-Vluyn: Neukirchener Verlagsgesellschaft, 2006, 137–165; Gottfried Hammann, Die Geschichte der christlichen Diakonie. Praktizierte Nächstenliebe von der Antike bis zur Reformationszeit, Göttingen: Vandenhoeck & Ruprecht, 2003.

**33** **1Tim. 6:17–19**: "Charge them that are rich in this world, that they be not high minded, nor trust in uncertain riches, but in the living God, who gives us richly all things to enjoy; That they do good, that they be rich in good works, ready to distribute, willing to communicate; Laying up in store for themselves a good foundation against the time to come, that they may lay hold on eternal life."

**34** **Rom. 3:21–24**: "But now the righteousness of God without the law is manifested, being witnessed by the law and the prophets; Even the righteousness of God which is by faith of Jesus Christ to all and on all them that believe: for there is no difference: For all have sinned, and come short of the glory of God; Being justified freely by his grace through the redemption that is in Christ Jesus."

**35** **Rom. 4:20–25**: Abraham "staggered not at the promise of God through unbelief; but was strong in faith, giving glory to God. And being fully persuaded that, what he had promised, he was able also to perform. And therefore it was imputed to him for righteousness. Now it was not written for his sake alone, that it was imputed to him; But for us also, to whom it shall be

order with just laws in society (Isa. 1:21–27),[36] since the Lord is not a God of confusion but of peace (1Cor. 14:33),[37] not a God of twilight and darkness but of light (Isa. 58:8–10).[38] Therefore, the aspiration of Psalm 85:10, that righteousness and peace will kiss each other, is the foundation of Christian love, realized through the reconciliation. This is furthermore the reality for Christians participating in the salvation deeds of Christ, and the aim for the social engagement of Christians.

This social commitment of Christians aiming for peace and fair conditions, however, is only one side of the coin. On the other side is the following announcement of Jesus that the Christian faith will split familiar and social ties:

> Suppose you that I am come to give peace on earth? I tell you, No; but rather division: For from now on there shall be five in one house divided, three against two, and two against three. The father shall be divided against the son, and the son against the father; the mother against the daughter, and the daughter against the mother; the mother in law against her daughter in law, and the daughter in law against her mother in law.
>
> (Luke 12:51–53)

Christians believe Jesus to be the Messiah and the Son of God; he brings forgiveness of sins and salvation. Presupposing these assertions to be true, Christians refute other religious or philosophical argumentations – and that causes division. If Jesus really is the Messiah, the Christians will not wait for him to come in the future; instead they will await his second coming on the Last Day. Moreover, the Christians trust in Jesus Christ and his proclaiming of the gospel

imputed, if we believe on him that raised up Jesus our Lord from the dead; who was delivered for our offenses, and was raised again for our justification."

**36 Isa. 1:21–27**: "How is the faithful city become an harlot! it was full of judgment; righteousness lodged in it; but now murderers. Your silver is become dross, your wine mixed with water: Your princes are rebellious, and companions of thieves: every one loves gifts, and follows after rewards: they judge not the fatherless, neither does the cause of the widow come to them. Therefore said the LORD, the LORD of hosts, the mighty One of Israel, Ah, I will ease me of my adversaries, and avenge me of my enemies: And I will turn my hand on you, and purely purge away your dross, and take away all your tin: And I will restore your judges as at the first, and your counsellors as at the beginning: afterward you shall be called, The city of righteousness, the faithful city. Zion shall be redeemed with judgment, and her converts with righteousness."

**37 1Cor. 14:33**: "For God is not the author of confusion, but of peace, as in all churches of the saints."

**38 Isa. 58:8–10**: "Then shall your light break forth as the morning, and your health shall spring forth speedily: and your righteousness shall go before you; the glory of the LORD shall be your rear guard. Then shall you call, and the LORD shall answer; you shall cry, and he shall say, Here I am. If you take away from the middle of you the yoke, the putting forth of the finger, and speaking vanity, and if you draw out your soul to the hungry, and satisfy the afflicted soul; then shall your light rise in obscurity, and your darkness be as the noon day."

(εὐαγγέλιον / evangelion). They build their life on repentance and faith and deny that good deeds are necessary for salvation and therefore to pass Judgement Day. Consequently, the division does not only touch cognitive claims of truth but also emotional settings of fidelity and obedience. To confess Jesus as the Christ has two meanings. It implies questioning the political and religious rulers (Luke 1:51f),[39] which will breed discord. Christians spread the gospel and missionized it in their surroundings according to their cognitive and emotional convictions. From the beginning, they experienced that their confessions were not shared by everyone, rather they lead to quarrels, and even ruptures in families, between friends, and in communities.

Hence the question arises, of how to live together in a society peacefully with these divergent convictions and how to tolerate them. In the beginning of Christianity these questions were not urgent, since the first congregations were waiting intensively for the Last Day to come very soon. Governed by this eschatological expectation, the first Christians were not in need to develop a constructive conflict culture; they lived a quiet and secluded life (Luke 21:25–36),[40] although there were conflicts, for example, between Paul and Peter (Gal. 2). However, since the Last Day was delayed, the necessity to accept living on this earth in one's lifetime grew. In the theological perspective, the experience not to find affirmation but repudiation was now interpreted by referring to the Holy Spirit. It is God himself who convinces humans or does not. Trusting in Jesus Christ is a spiritual gift, and not the efficacy of humans (1Cor. 12:1–3).[41] In society, this problem

**39 Luke 1:51–52**: God "has showed strength with his arm; he has scattered the proud in the imagination of their hearts. He has put down the mighty from their seats, and exalted them of low degree."

**40 Luke 21:25–36**: "And there shall be signs in the sun, and in the moon, and in the stars; and on the earth distress of nations, with perplexity; the sea and the waves roaring; Men's hearts failing them for fear, and for looking after those things which are coming on the earth: for the powers of heaven shall be shaken. And then shall they see the Son of man coming in a cloud with power and great glory. And when these things begin to come to pass, then look up, and lift up your heads; for your redemption draws near. And he spoke to them a parable; Behold the fig tree, and all the trees; When they now shoot forth, you see and know of your own selves that summer is now near at hand. So likewise you, when you see these things come to pass, know you that the kingdom of God is near at hand. Truly I say to you, This generation shall not pass away, till all be fulfilled. Heaven and earth shall pass away: but my words shall not pass away. And take heed to yourselves, lest at any time your hearts be overcharged with surfeiting, and drunkenness, and cares of this life, and so that day come on you unawares. For as a snare shall it come on all them that dwell on the face of the whole earth. Watch you therefore, and pray always, that you may be accounted worthy to escape all these things that shall come to pass, and to stand before the Son of man."

**41 1Cor. 12:1–3**: "Now concerning spiritual gifts, brothers, I would not have you ignorant. You know that you were Gentiles, carried away to these dumb idols, even as you were led. Why I

is not yet solved, although Christendom has developed a compromise that time and again has mitigated it – as I will explain further in the second section.

## 1.4 From the Political Perspective

The apostle Matthew gave another wording of Jesus' announcement in his gospel: "Think not that I am come to send peace on earth: I came not to send peace, but a sword" (Matt. 10:34). The division caused by Jesus and his claim to announce the truth about God and to establish faith as the one way to God is labelled here with the term sword. On the one hand, sword is a metaphor for the resoluteness of faith, it cuts between things that belong to the Christian faith and others, that do not. On the other hand, albeit, this metaphorical way of speaking exacerbates the social conflict to a political struggle; the sword represents the state's use of violent force. Soldiers and warriors as the executive power of the government wear a sword and are authorized and trained to use it. Did Jesus announce himself as and his message in order to achieve political changes? We know that one of his disciples was a Zealot (Luke 6:15: "Simon called Zelotes"), who wanted to defeat the Romans with violent means, including plots and assaults.[42] However, his order to a disciple in Gethsemane, who wanted to attack with a sword, clearly speaks against the assumption that Jesus wanted radical political changes using violent means:

> Then came they, and laid hands on Jesus and took him. And, behold, one of them which were with Jesus stretched out his hand, and drew his sword, and struck a servant of the high priest's, and smote off his ear. Then said Jesus to him, Put up again your sword into his place: for all they that take the sword shall perish with the sword. (Matt. 26:50–52)

Jesus did neither take up the sword nor allowed his disciples to fight with weapons. The Kingdom of God he proclaimed has religious and spiritual, but not political aims (John 18:36).[43] For that reason his disciples must not fight for it with military means. The peace with God cannot be achieved with the sword,

give you to understand, that no man speaking by the Spirit of God calls Jesus accursed: and that no man can say that Jesus is the Lord, but by the Holy Ghost."

**42** Cf. Hengel, Martin, *Die Zeloten. Untersuchungen zur jüdischen Freiheitsbewegung in der Zeit von Herodes I. bis 70 n. Chr.*, Leiden/Köln: E.J. Brill, 1976, 72–76, 400.

**43** **John 18:36**: "Jesus answered: My kingdom is not of this world: if my kingdom were of this world, then would my servants fight, that I should not be delivered to the Jews: but now is my kingdom not from hence."

but rather with repentance and faith. It does not matter why Matthew has picked up the term sword in this context (Matt. 10:34); Christians should be cautious to pick up his wording in any case. They should rather refer to Luke's version of this announcement (Luke 12:51).

Nevertheless, Matthew's wording points to a challenge for the Christians to meet: How shall Christians relate to the political leaders? The relationship between politics and religion was always a central concern for religion. Both terms denote powers that rule over humans and intervene in their life gravely: kings and priests. We are dealing with two alpha leaders, so to speak. The basic question is how to relate faith and political obedience. Is it possible to balance the religious and the political call for loyalty and devotion? The New Testament gives three answers to this question.

The first answer is the concept of coexistence: According to Matthew, Jesus has already been confronted with this question, and he answered to the Pharisees: "Render therefore to Caesar the things that are Caesar's, and to God the things that are God's" (Matt. 22:21). More precisely he was talking about taxes, which are secular affairs, and thus irrelevant for God's Kingdom. In heaven no one will have to pay taxes. Finally, God's Kingdom is not from this world (John 18:36); it is not about eating and drinking, but. about righteousness, peace, and joy in the Holy Spirit (Rom. 14:17). The secular forces are just there as many other things in the world are, but nothing more. Thus, Christians should not deny to obey those rulers, they should fulfill their secular duties, but with a clear and distinct prioritization. As long as these duties do not affect the faith or God's Kingdom, the Christians are allowed to fulfill them alongside their obedience of faith. However, whenever there is an ambiguity or, even worse, an abuse of secular powers, Peter's Clausula sets the agenda: "We ought to obey God rather than men" (Acts 5:29).

The second answer is the concept of contradiction: In the Old and also in the New Testament one will find apocalyptic writings that refer to an eschatological battle between God and the forces of evil. These forces will attack not only the Church but the whole of creation. In the background the experienced persecution and bullying of the pious are reflected theologically: Even the worst events will be overruled by the biblical God. In this conception the secular forces do not coexist with God, but contradict and combat him and his congregations. The prophet Daniel portrayed the evil forces as monsters in the Hebrew Bible, which symbolizes the most vicious powers, and their violence would accumulate up to the bloody end (Deut. 7). The persecuted Christians have only one opportunity: to wait for God's intervention without losing their faith (James 1:12).[44] There will be a happy ending! God will surely step in and end the bad actions. On Doomsday he will punish the evildoers, redeem and reward the faithful, and hereafter the age of eternal happiness will begin (Luke 21:28). Consequently, the Christians shall draw back from this world (Luke 21:21); they shall not combat the evil forces, because only God can defeat the devil. Instead, they shall follow Jesus in suffering the pain and furthermore use their apocalyptic knowledge to strengthen their resilience.[45]

The third answer is the concept of cooperation: When the apostle Paul wrote his letter to the Romans to introduce himself and his planned mission in Spain, he elucidated his concept of relationship between faith and secular powers to the politically aware citizens in the center of the empire:

> Let every soul be subject to the higher powers. For there is no power but of God: the powers that be are ordained of God. Whoever therefore resists the power, resists the ordinance of God: and they that resist shall receive to themselves damnation, for rulers are not a terror to good works, but to the evil. Will you then not be afraid of the power? Do that which is good, and you shall have praise of the same: For he is the minister of God to you for good. But if you do that which is evil, be afraid; for he bears not the sword in vain: for he is the minister of God, a revenger to execute wrath on him that does evil. Why you must needs be subject, not only for wrath, but also for conscience sake. For this cause pay you tribute also: for they are God's ministers, attending continually on this very thing. Render therefore to all their dues: tribute to whom tribute is due; custom to whom custom; fear to whom fear; honor to whom honor. (Rom. 13:1–7)

These remarks frame the coexistence Jesus had taught the Pharisees with theological arguments. The secular rulers in Matthew's quotation have been modified in Paul's diction, he now names them a state, which furthermore has authority and not only power, and of course this authority arises from God. Consequently, the Christians shall not only accept the coexistence with this state but may cooperate, since the political authorities are assigned by God. To obey the secular authorities now implies to accept their divine commission (Titus 3:1–2)[46]

---

**44 James 1:12**: "Blessed is the man that endures temptation: for when he is tried, he shall receive the crown of life, which the Lord has promised to them that love him."

**45** Cf. Scherer, Hildegard, "Gewalt bewältigen. Neutestamentliche Stimmen," in: Jochen Flebbe/ Görge K. Hasselhoff (eds), *Ich bin nicht gekommen, Frieden zu bringen, sondern das Schwert. Aspekte des Verhältnisses von Religion und Gewalt*, 69–90, Göttingen: Vandenhoeck & Ruprecht, 2017.

**46 Titus 3:1–2**: "Put them in mind to be subject to principalities and powers, to obey magistrates, to be ready to every good work, to speak evil of no man, to be no brawlers, but gentle, showing all meekness to all men."

and even to include them into the Christians prayers (1Tim. 2:1–2).[47]– Whether this obedience also includes the willingness of Christians to engage in politics cannot be decided on the basis of the biblical sources. Regardless it would be consistent with the progression of this concept of cooperation and a clear distinction from the concept of contradiction.

In summary, the relationship between faith and secular powers depends on one's perspective of the world. It depends on whether the world is thought of as a secular and temporary fact coexisting with God, or as a threatening opponent to God that will be defeated in the end, or as an institution that is authorized to cooperate with God. Nevertheless, in all three cases, the biblical God defines the relationship. So the first commandment doubtlessly has the priority. Consequently, the peace with God is still the primary aim. God will destroy the evildoers, personally as well as politically (Ps. 92:8–10[48]). In the end he will even destroy death, his last enemy (1Cor. 15:26). Yet for now the political authorities are also accepted as a force that supports God's aim by fighting and limiting the evil on earth and by this making their contribution to establish at least a negative peace.

## 2 Modifications and Developments of Peace in Church History

Christianity has expanded from the first congregations departing from the Jewish context to a world-wide religion. Obviously this development cannot be represented in this essay. Instead I will highlight a few positions and arguments that have greatly influenced the Christian understanding of peace. Following the systematic approach, I will not represent them chronologically, I will focus on the four perspectives instead.

47 **1Tim. 2:1–2**: "I exhort therefore, that, first of all, supplications, prayers, intercessions, and giving of thanks, be made for all men; For kings, and for all that are in authority; that we may lead a quiet and peaceable life in all godliness and honesty."

48 **Ps. 92:8–10**: "But you, LORD, are most high for ever more. For, see, your enemies, O LORD, for, see, your enemies shall perish; all the workers of iniquity shall be scattered. But my horn shall you exalt like the horn of an unicorn: I shall be anointed with fresh oil."

## 2.1 From the Political Perspective

In the Early Church the three concepts of relationship between religion and politics coexisted. Some parishes, experiencing a persecution of Christians, tended to the apocalyptical concept of contradiction, whereas other parishes could coexist with the Romans and their religious and philosophical mindset. Despite the sparse sources of these times we know of some Christians working as soldiers (Luke 3:14;[49] Acts 10: the centurion Cornelius becomes a Christian),[50] furthermore, there are the diaconal deeds of the Christians; therefore we can also speak of cooperation. To classify these notions, one must nevertheless keep in mind that the parishes were a minority in the beginning.

In the fourth century AD, radical changes took place. Whilst at the beginning of Christianity there were official and wide-spread persecutions of Christians, the political atmosphere suddenly changed: the Constantinian Shift. The Roman Emperor Constantine became a Christian and from that time he supported the Churches.[51] First he legalized Christianity in the year 313 (Edict of Milan). Then he supported the council of Nicaea (325), an attempt to end the debates about the understanding of the trinity and to formulate a binding creed for all Christians. After his conversion, Constantine avoided participation in pagan rituals, and towards the end of his life, he was baptized. Yet paganism and Christianity coexisted until the Emperor Theodosius declared Christianity to be the official religion of the Roman Empire in the year 381 – now prohibiting the "heathen" (pagan) rituals.

Focusing on the understanding of peace in the political perspective, the ongoing debates about the character of the Emperor's faith and the complex consequences for Christianity cannot be considered here. It is important, however, that Constantine's conversion was combined with his military efforts. According to the Church historian Eusebius (263–339), Constantine had a vision of a cross of light combined with the words "*in hoc signo vinces*" (in this sign you will conquer)

49 **Luke 3:14**: "And the soldiers likewise demanded of him [= John the Baptist], saying, And what shall we do? And he said to them: Do violence to no man, neither accuse any falsely; and be content with your wages."

50 Cf. Markschies, Christoph, *Das antike Christentum. Frömmigkeit, Lebensformen, Institutionen*, München: C.H. Beck, 2016.

51 Cf. van Dam, Raymond, *The Roman Revolution of Constantine*, Cambridge: Cambridge University Press, 2007; Girardet, Klaus Martin, *Die konstantinische Wende. Voraussetzungen und geistige Grundlagen der Religionspolitik Konstantins des Großen*, Darmstadt: Wissenschaftliche Buchgesellschaft, 2006.

before the Battle of the Milvian Bridge (312).[52] So he added the symbol of the cross to the vexillum (the flag of the army), convinced that he would win the battle under this sign. And he did indeed win. Through this act the biblical God was identified with the "*Sol invictus*," the Roman sun–god and the patron of the soldiers. Furthermore, war as a political means was now combined with the cross of Jesus Christ. Admittedly in the Hebrew Bible there are some passages that also combine God with war. God supports his chosen people against their enemies using natural forces (Josh. 10:5–11)[53] and trembling in the host (1Sam. 14:15).[54] In the Hebrew Bible, however, these narratives are mainly found in the ancient history of Israel. In contrast, later passages such as Psalm 85 stress the combination of God with righteousness and peace – as previously discussed presented. Especially Jesus' repudiation of the sword and his blessing of the peacemakers do not fit with Constantine's interpretation of the cross as a symbol for military victory. Has war by now become an accepted means for Christianity? Or further, was war now a divine instrument used by the emperor to defeat God's enemies?

The Church Father Augustine (354–430) was the first to answer these questions with his so called doctrine of the two kingdoms and with his just war theory.[55] He was theologically challenged by the Sack of Rome by the Visigoths (410). On one hand, the pagan Romans interpreted this event as revenge from the

---

52 Cf. Amerise, Marilena, "Monotheism and the Monarchy. The Christian Emperor and the Cult of the Sun in Eusebius of Caesarea," *Jahrbuch für Antike und Christentum* 50 (2007), 72–84.

53 **Josh. 10:5–11**: "Therefore the five kings of the Amorites, the king of Jerusalem, the king of Hebron, the king of Jarmuth, the king of Lachish, the king of Eglon, gathered themselves together, and went up, they and all their hosts, and encamped before Gibeon, and made war against it. And the men of Gibeon sent to Joshua to the camp to Gilgal, saying, Slack not your hand from your servants; come up to us quickly, and save us, and help us: for all the kings of the Amorites that dwell in the mountains are gathered together against us. So Joshua ascended from Gilgal, he, and all the people of war with him, and all the mighty men of valor. And the LORD said to Joshua, Fear them not: for I have delivered them into your hand; there shall not a man of them stand before you. Joshua therefore came to them suddenly, and went up from Gilgal all night. And the LORD discomfited them before Israel, and slew them with a great slaughter at Gibeon, and chased them along the way that goes up to Bethhoron, and smote them to Azekah, and to Makkedah. And it came to pass, as they fled from before Israel, and were in the going down to Bethhoron, that the LORD cast down great stones from heaven on them to Azekah, and they died: they were more which died with hailstones than they whom the children of Israel slew with the sword."

54 **1Sam. 14:15**: "And there was trembling in the host, in the field, and among all the people: the garrison, and the spoilers, they also trembled, and the earth quaked: so it was a very great trembling."

55 Cf. for the following Flasch, Kurt, *Augustin. Einführung in sein Denken*, Stuttgart: Reclam, [2]1994, and Weissenberg, Timo J., *Die Friedenslehre des Augustinus. Theologische Grundlagen und ethische Entfaltung*, Stuttgart: Kohlhammer, 2005.

ancient gods for the Romans' perfidy of converting to Christianity and ignoring the traditional rituals. Accordingly, Augustine had to demonstrate that this pillage was not connected with the Christian faith. On the other hand, it seemed necessary to defend the Church and the Christendom with military means against the attack from the North. Thus, he had to resolve the acceptance of military means from the Christian point of view.

To solve these queries, Augustine distinguishes in his 22 books "De Civitate Dei" (426) God's kingdom as the City of God from the earthly kingdoms as the City of Man. The City of God is not from this world (John 18:36). Consequently, eternal peace will be realized in heaven, not on earth. A person believing in Jesus Christ is counted to the chosen ones, and will come to heaven after his death. Thus, the wars on earth as all mundane events have nothing to do with the City of God, but only with the City of Man. Therefore, neither the Church nor the Christian faith is responsible for the current military failure. Augustine refutes by this an apocalyptic interpretation of history; the Roman Empire is not the end of history. On earth, in the City of Man there is an ongoing struggle between God and the devil since the Fall of Man and Cain's fratricide (Gen. 3–4). The Church and the political government shall support God and compete against the devil and his forces with all means. Violent means and military power may also be used to limit sins, oppose the heretics and defeat the offenders from the North. The aim is to establish or preserve the public order and tranquility so that the Church can preach the gospel and call the people up to the transcendent City of God. With this doctrine, Augustine combines the biblical concepts of contradiction and coexistence of Church and state. The City of God is a contradiction to the City of Man; it is not from this world. Nevertheless, during this aeon the government will coexist with the Church and can furthermore subserve the Church by ensuring good conditions for preaching the gospel.

In Mediaeval Times this doctrine faded into the background, because Church and government were widely interconnected – more about this in the next subsection. The reformer Martin Luther (1483–1546), however, a monk from the Augustinian Hermits fraternity, took this doctrine up again and modified it to his doctrine of the two governments.[56] He wrote:

> For this reason God has ordained two governments: the spiritual, by which the Holy Spirit produces Christians and righteous people under Christ; and the temporal, which

56 Cf. for the following Stümke, Volker, *Das Friedensverständnis Martin Luthers. Grundlagen und Anwendungsbereiche seiner politischen Ethik*, Stuttgart: Kohlhammer, 2007.

> restrains the un-Christian and wicked so that – no thanks to them – they are obliged to keep still and to maintain an outward peace.[57]

Church and state have the duty to look after the people on earth. One is responsible for human salvation, the other for peace on earth. As Augustine, Luther is also convinced that God's Kingdom is transcendent and must not be admixed with political empires. Furthermore, both conform that on earth there is a struggle for the souls of humans between God and the evil forces. For Luther, though, Church and government do not only coexist, but cooperate. Both are ordained by God, both shall serve him with their specific instruments and perform their respective duties. The Church is geared to spiritual means; it shall preach the gospel and by this lead humans to become Christians and to gain the City of God. The government is, according to Rom. 13, also ordained by God. The temporal regiment governs the people and is governed by God. God does not directly interfere in political affairs but assigns the ruler with authority, endows him or her with a distinct function and equips the ruler with helpful instruments as the sword and the laws. Moreover, the aims for the government are also modified by Luther. Not order and tranquility, but peace and justice are the divine terms of reference.[58] The government has the duty to maintain negative peace on earth and it can use even military means (the sword) to achieve this goal.

To sum up: In the Christian tradition it was widely accepted that the government is ordained by God with the aim to preserve the negative peace on earth. This peace is, following Augustine, a condition to preach the gospel. Whereas the evil forces favor chaos and confusion, the Church needs order and tranquility, only under these circumstances can the Christians proclaim the gospel and the humans listen to it. According to Luther, negative peace is an end in itself. It is God's will that people shall live together without war and combat. Both agree that in order to establish negative peace the government can use the sword. Whereas the Church, according to Matt. 26:52 is not allowed to use violent means, the state has the monopoly on the legitimate use of force, and that even includes the allowance of military action.

---

**57** Luther, Martin, *Temporal Authority. To What Extend it Should be Obeyed* (1523), LW 45, 91 =WA 11, 251. Further: Leonhardt, Rochus /von Scheliha, Arnulf (eds), *Hier stehe ich, ich kann nicht anders! Zu Martin Luthers Staatsverständnis*, Baden-Baden: Nomos, 2015.

**58** Cf. Luther, Martin, *Whether Soldiers, too, Can be Saved* (1526), LW 46, 96 = WA 19, 625: "For the very fact that the sword has been instituted by God to punish the evil, protect the good, and preserve peace [Rom. 13:1–4; 1Pet 2:13–14] is powerful and sufficient proof that war and killing along with all the things that accompany wartime and martial law have been instituted by God."

This acceptance of forceful means is, however, ambivalent, or even unacceptable, for Christians, if the violence is not limited. The biblical experiences that led to the concept of contradiction must not be overruled by this comprehensive allowance of violence and even war. At this point, Augustine picked up the theory of just wars from the Roman philosopher Cicero to limit the acceptance of armed encounters. According to Augustine, war is primarily a state of disorder and therefore strengthens the evil forces. The combatants are driven by greed and vengeance; these inferior internal forces dominate the virtues as love and righteousness and so the soul of the soldier is in disorder. This disorder must be conquered, however, in order to regain order and tranquility, and this may only be available through military violence. Consequently, these wars are simply an aim to end the disorder and to establish negative peace on earth; and they are named "just wars." Albeit, this legitimate aim is not sufficient for a just war yet; in addition, there must be a good reason to start a war, for example a state of disorder that was caused by the enemy. Furthermore, the process of declaring war must be correct. Not just anyone, but rather only God or the legal political leader is allowed to use violent means, otherwise disorder will arise in the government as well. Finally, all these conditions must not contradict God's commandments.

Augustine's intention to limit wars by developing ethical criteria for a just war was continued in the mediaeval scholasticism.[59] Very important was the conception from Thomas Aquinas (1225–1274).[60] According to Thomas, only those acts of violence that correspond to the criteria of just war are reconcilable with Christian faith. And of course, these were the only acts of violence that were justified to be exercised through the Holy Roman Empire of the German Nation. More precisely the Doctor of Christianity postulated three criteria:

- legitimate authority (*legitima potestas*[61]),
- just cause (*causa iusta*),
- right intention (*recta intentio*).

---

59 Cf. for the following the anthology edited by Justenhoven, Heinz-Gerhard/Barbieri Jr. William A., *From Just War to Modern Peace Ethics*, Berlin/ Boston: de Gruyter, 2012.

60 Cf. Beestermöller, Gerhard, *Thomas von Aquin und der gerechte Krieg. Friedensethik im theologischen Kontext der Summa Theologiae*, Köln: J.P. Bachem, 1990.

61 Thomas Aquinas did not speak of the legitime authority but more specific of the authority of the prince resp. the ruler (auctoritas principis). Since the Spanish Late Scholasticism the criterion was modified so that not the person (e.g. the Pope or the Prince) but the government was foregrounded. Besides, this allowed including different forms of regimes and so better fits to the Modern Ages.

By this, civil wars, relish wars and depredations were forbidden, whereas defensive wars and punitive actions (as an aggressive war) were allowed. In the late Middle Ages more criteria for the legitimation of wars (*ius ad bellum*) were added, notably the condition of the last resort (*ultima ratio*). Furthermore, not only the war, but the conduct in battle was analyzed as well. The most important criteria of this "*ius in bello*" were the discrimination between combatants and non-combatants, the proportionality of means (*debitus modus*) and the limitation of collateral damages (doctrine of double effect).

These just war theory criteria, with only slight modifications, are largely uncontested and still applicable today. They are still an effective instrument to evaluate wars and military use of violence.[62] It is for this reason that, especially in the English-speaking world, many theologians still adhere to the just war theory.[63] There is also an ongoing discourse on the just war theory in philosophy and in the political sciences.[64] The intention to establish ethical criteria for just wars and thereby limit wars also remains uncontested.[65] All philosophers and theologians in the current debates are taking up this doctrine in order to restrict wars according to an ethical benchmark.

The implicit acceptance of war as a political tool is, however, contested. In the just war tradition, even aggressive wars could be legitimized, if there was a just cause (preceding wrongs), a right intention (punishment) and a correct declaration (through the emperor). Luther whereas reduced his acceptance to defensive wars. Yet, in both cases, does political realism not overrule the Christian hope for peace? Is Christian faith not limited to the personal wish for eternal life anyway? It is one thing accepting the government as one of the two regiments God has ordained, and accepting the divine aims, including the government's

62 The memorandum of the EKD from 2007 (*Aus Gottes Frieden leben – für gerechten Frieden sorgen. Eine Denkschrift des Rates der Evangelischen Kirche in Deutschland*, Gütersloh: Gütersloher Verlagshaus, 2007) for example makes recourse in paragraph 102 to the following criteria, now renamed criteria of right-preserving force (Kriterien rechtserhaltender Gewalt): the right cause, legitimate authority, right intention, last resort, proportionality of consequences and means and the discrimination principle. Albeit, these criteria do not legitimate a war, but they label measures to preserve or to rebuild a legal system or a state law.

63 Cf. Biggar, Nigel, *In Defence of War*, Oxford: Oxford University Press, 2013; McMahan, Jeff, *Killing in War*, Oxford: Oxford University Press, 2009.

64 Cf. Frowe, Helen, *The Ethics of War and Peace. An Introduction*, London: Routledge, $^{2}$2016; Rodin, David/ Shue, Henry (eds), *Just and Unjust Warriors. The Moral and Legal Status of Soldiers*, Oxford: Oxford University Press, 2010.

65 Cf. Bartolomé de Las Casas, who stated, that there was not even one just war in the combats of Spain against the Indians; in: Gillner, Matthias, "Bartolomé de Las Casas und die Menschenrechte," *Jahrbuch für christliche Sozialwissenschaften* 39 (1998), 143–60, 152.

monopoly of the legitimate use of force. Accepting war as a natural component of this use of force is another. Such acceptance does not take adequately into account that Jesus, as the Christ, is our peace (Eph. 2:14) and that his salvation will also reach and change the society. It is necessary therefore, to complement the political point of view with the social perspective on the rule of law.

## 2.2 From the Social Perspective

After Christianity had become the official religion of the Roman state in the year 381 the relationship between Church and state changed. The political implications were just discussed; the government had become part of Christendom, and political means up to the use of military force were accepted by the Church. This concept of cooperation, however, also modified the social position of the Church. As the official representative of the Christian faith the Church, and especially the Pope, became more important and influential. This development was bolstered with another saying of Jesus. In the garden of Gethsemane, when Jesus was captured, his disciples said: "Lord, behold, here are two swords. And he said to them, It is enough" (Luke 22:38). These two swords became the symbol for the two powers in the Roman and later the Christian society.

During history of the Church different terms were used to denote the two forces in society: *civitates* (cities), *potestates* (powers), *gladii* (swords), and *regimina* (governments).[66] Each of these terms grammatically denotes that there was exactly two of each of them. Albeit it is disputed who has the final word, meaning who has supremacy in society, Church or state. On one hand there were the Franconian emperors with their proprietary Church system, implying that the Churches belong to the particular rulers. According to this concept, the state would have the final say and the Church would become the department for religion in each princedom. The Church protested against this concept, because freedom of speech, or more precisely the freedom to proclaim the Word of God, would be restricted. So on the other hand the Church established a counterproposal. It argues that the secular emperor is ordained by the Pope and receives orders on ruling the people. In this concept, however, the political actions are too limited. In summary, the distribution of power between Church and state seemed to be impossible; both wanted to dominate over the other.

66 Cf. Stümke, Volker, *Das Friedensverständnis Martin Luthers*, 196–199. „Regimina“ is the latin translation from Luthers „Regimente“.

In the Medieval Ages, a compromise was established. The Concordat of Worms (1122 AD) laid down that Church and state, Pope and Emperor, have to manage different tasks and duties and by this shall respect each other. This compromise was helpful to solve the problem concretely with the investiture of the bishops. The Concordat distinguished between the *temporalia* and the *spiritualia* and by this found a solution. Since the Church is responsible for the spiritual powers of the bishop, such as administration of the sacraments, and since these are the primary tasks for a bishop, he is invested by the Church. However, since there are also secular warrants of the bishop, such as to administer the Church properties (i.e. to lease and release land), this authorization is developed by the Emperor. This compromise leads to an accepted coexistence of Church and state with no one dominating the other. A department of religion and a religious state were excluded.

In the Orthodox Churches, the relationship between Church and state was and is thought of similarly as "symphonia" (*συμφωνία*), a harmonic cooperation between religious (Aaron) and political (Moses) leaders or between the body and the soul.[67] Whereas the Roman (Catholic and Protestant) Churches tended to divide the two forces, the Orthodox Churches were and are more closely linked with the government; most of them are national Churches, for example the Russian Orthodox Church and the Greek Orthodox Church. Consequently, up to now they do not agree with the concept of just peace; instead understand peace service as an assignment for the local Churches.

In Western Christianity, the influence of the Church and the Pope was manifested in the Concordat of Worms. Nevertheless, the term "sword" still implies two challenges for Christendom. On the one hand, the Church was combined with military means. Obviously it was metaphorical speech; the spiritual sword shall not kill people, but defeat sins. However, the crusades demonstrate that a combination was not out of sight: to kill the infidels with both swords. On the other hand, the Christians are still bound to Matthew 26:52 and are consequently prohibited from using the sword. Certainly, the Church could delegate the use of force to the state, nevertheless, the politicians and soldiers are Christians as well. Therefore, this solution tends to a grading of Christians, to a step range between perfect (pure) and imperfect (impure) Christians, the first avoid taking up a sword, the second use violent means. This grading, however, tends to contradict Paul's doctrine of justification: "For you are all the children of

67 Cf. Oeldemann, Johannes, *Die Kirchen des christlichen Ostens. Orthodoxe, orientalische und mit Rom unierte Kirchen*, Düsseldorf: Topos, 2016; McGuckin, John Anthony, *The Orthodox Church. An Introduction to Its History, Doctrine, and Spiritual Culture*, Oxford: John Wiley and Sons, 2010.

God by faith in Christ Jesus. For as many of you as have been baptized into Christ have put on Christ. There is neither Jew nor Greek, there is neither bound nor free, there is neither male nor female: for you are all one in Christ Jesus" (Gal. 3:26–28).

It was again Martin Luther who developed a line of thought to solve these challenges. In his doctrine of the two governments, he renamed the spiritual means replacing "sword" by "word". Only the government, not the Church, must handle the sword. The Church has to convince mankind "*sine vi humana, sed verbo*"[68] (without violence, with words instead). The Church has to proclaim the divine word with human words, meaning that it has to preach the gospel, to call for repentance and faith, and to make use solely of the means of oration and its possibilities. A crusade is in any case not an option for Christians. Neither is the Church allowed to use the sword herself, nor is it permitted to call on the state to combat for religious reasons, for *spiritualia*. In the first case the Church would use a mean without authorization. In the second case it would urge the government to intervene in *spiritualia* with earthly, temporal means.

With his wording Luther does not only define more precisely the duties of the Church, he also stresses the state's monopoly on the use of force. Church and state have different duties in society and can draw on different means to fulfill them. One is responsible for human salvation, the other for earthly peace. This distinction picks up the biblical concept of coexistence and enhances it with the argument that both institutions are installed by God and because of this shall cooperate to look after the temporary and spiritual health of humans in society. Furthermore, since both governments are directly ordained by God, the hierarchical struggles between these two alpha leaders come down to nothing. Both are ordained by God and are therefore responsible to God. Accordingly, they are thus not allowed to shape their own targets arbitrarily. Hence, neither a totalitarian regime nor an absolute government is permitted, since both regimes restrict each other and, furthermore, are both limited through God.

Thus, Luther's doctrine of the two governments is still relevant today to solve the first challenge. In society, the Christian Church does not have the duty of handling the sword. It is not responsible for politics, but for religion instead, including e.g. preaching, pastoring, teaching, and deaconship. With these means it may lead humans to the Christian faith and by this to a content

**68** "Confessio Augustana Art. 28," in: *Die Bekenntnisschriften der evangelisch-lutherischen Kirche* [BSELK], Göttingen: Vandenhoeck & Ruprecht, [8]1979, 124, 21 – for Luther cf. WA 11, 268f (Temporal Authority: To What Extent it Should be Obeyed, 1523) and WA 32, 150ff (Sermon on Eph. 6,10ff from 11. November 1530).

life of peace with God. Accordingly, peace on earth, more precisely the negative peace is the temporal aim for the secular government. The state is ordained by God to stop wars and combat and to establish a legal order with violent means. More than this negative peace cannot be achieved by politics, since it cannot rule the conscience of humans. Thoughts are free, and the conscience can only be convinced by words, as I will explain further in the next subsection.

The second challenge, however, is not yet solved. On the one hand, the state has the duty to use violent means through its officers. On the other hand, Christians shall not pick up the sword. Thus, can a Christian become a soldier or a police officer? Or is it better for him or her to keep distance from politics and to lead a calm and tranquil life, for example in a monastery? Luther protested against such a grading. Referring to Paul, and according to his doctrine of justification, being a Christian depends solely on God's grace and Christian faith, but never on deeds. And to live in a monastery is for Luther a human deed. So it is not the profession but the confession that marks a Christian. Nevertheless, there are evil deeds and professions that should be avoided by Christians, murder for example. At this point, Luther picks up the doctrine of the two governments. Since the state is ordained by God and equipped with his specific means, Christians are not only allowed but told to support the secular government with their professions. Following Luther, a Christian may become a soldier or even an executioner, since he or she supports the temporary inputs of God with this profession – like a teacher or a priest does. Moreover, Luther even combines these professions with the neighborly love of Christians. If one takes into account that there can be more than one other person to deal with, it becomes evident, that not the offender but the potential victim is the neighbor that must be protected by the Christian bystander. Protecting the victims is an expression of Christian love, although it might hurt the perpetrator. This harmonizes with Luther's limitation on defensive wars: The sword may be used exclusively to protect humans in society and this aim can also be supported by Christians.

Nevertheless, Christian faith is still combined with violent means. Thus, not all Christians have accepted this line of argumentation.[69] In the Middle Ages, the Cathari and the Waldensian refused any kind of military service.[70] In

**69** Cf. Werkner, Ines-Jacqueline, "Kirchliche Diskurse um die Anwendung militärischer Gewalt. Eine empirische Perspektive," in: Sarah Jäger/Ines-Jacqueline Werkner (eds), *Gewalt in der Bibel und in kirchlichen Traditionen. Fragen zur Gewalt Band 1*, 87–116, Wiesbaden: Springer, 2018.
**70** Cf. Lambert, Malcolm, *Geschichte der Katharer*, Darmstadt: Wissenschaftliche Buchgesellschaft, 2001; Audisio, Gabriel, *Die Waldenser*, Augsburg: Bechtermünz, 2004.

the Anabaptist movement during the Reformation period the Hutterites and the Mennonites likewise developed a stringent peace-ethic with an abolition of the use of force by Christians.[71] Since then, the Quakers and the Adventists also belong to these Peace Churches.[72] All these Christians have in common that they take the appeal of the Sermon on the Mount (Matt. 5:38–42) very seriously and therefore reject any form of force, including service in military and police forces. According to these groups, Christians should symbolize peace with God in their daily life and also in their professions.[73] They should not be conformed to this world: but be transformed by the renewing of their mind; by this, they may prove what that good, and acceptable, and perfect, will of God is (Rom. 12:2).

Regarding the necessity of social engagement these Peace Churches differ though. Some withdraw from society into their communities to lead a tranquil and calm life. They pay their taxes and obey the laws according to Rom: 13, as long as they do not contradict the Clausula Petri (Acts 5:29), but they are not involved in political or social activities outside of their Church communities. However, by this they risk being questioned and criticized in that instead of supporting others they shirk their responsibility for protecting the weak. Luther put it plainly: It is demanded of a Christian to respond peacefully to an assault, as far as he is concerned only for himself or herself; but for the other's welfare, to protect them from harm, he must defend them. This is to be extended on society; Christians should therefore promote the authority of the state instead of withdrawing from social responsibility.

Other Christians in Peace Churches are, in contrast, fully committed to social welfare. They are involved in feeding the poor, engage in peace missions and promote peaceful means.[74] At the same time they criticize very convincingly the quite natural recourse to force as a legitimate political instrument: Peace on earth can only be established by peaceful means and Christians are called to follow Jesus, who convinced humans with his words and with his neighborly love and not with the sword. Does the vital pacifism of these Churches not demonstrate how to

---

**71** Packull, Werner O., *Hutterite Beginnings. Communitarian Experiments during the Reformation*, London: Johns Hopkins University Press, 1995; Lichdi, Diether G., *Die Mennoniten in Geschichte und Gegenwart. Von der Täuferbewegung zur weltweiten Freikirche*, Weisenheim: Agape, [2]2004.
**72** Cf. Henke, Manfred, *Wir haben nicht einen Bettler unter uns. Studien zur Sozialgeschichte der frühen Quäkerbewegung*, Berlin: be.bra wissenschaft verlag, 2015; Knight, George R., *Anticipating the Advent. A Brief History of Seventh-day Adventists*, Nampa: Pacific Pr Pub Assn, 1993.
**73** Cf. Enns, Fernando, *Friedenskirche in der Ökumene. Mennonitische Wurzeln einer Ethik der Gewaltfreiheit*, Göttingen: Vandenhoeck & Ruprecht, 2003.
**74** Cf. Hofheinz, Marco/van Oorschot, Frederike (eds), *Christlich-theologischer Pazifismus im 20. Jahrhundert*, Baden-Baden: Nomos, 2016, esp. 141f and 203.

overcome violence without shirking social responsibility? Consequently, the Catholic Church and the large Protestant Churches currently tend to incorporate these valid objections and suggestions in their conception of just peace – yet that will lead us to the next section.

In summary, it is currently controversial how Christians shall behave in society. Whereas it is evident that the Church is not allowed to use forceful means as the state, which is regulated by laws and has the monopoly on military force, the involvement of Christians in this political use of armed force is moot. On the one hand it is convincing that perpetrators must be hindered from hurting humans, and consequently, the state must be allowed to use violent means (Rom. 13). Should Christians not protect their neighbors as well? On the other hand, it is likewise plausible that Christians shall promulgate the peace with God and its consequences for society with their words and deeds (Matt. 5). Is it not more convincing for Christians following Jesus to react peacefully on an assault than to combat it and continue in the cycle of violence? Furthermore, we have to take into account that apart from the state and the Church there are currently additional global players in society, for example, the economy, sports, sciences, culture, and media. Two "swords" are today definitely not enough to paraphrase the influential forces in society. This opens more opportunities to react to violence using, for example, economic boycotts, diplomatic approaches, public naming and shaming, and social defiance. The instruments of armed force are the most destructive ones. Consequently, the state needs responsible and circumspect officers to limit the use of force as far as possible. Can Christians reject this need of cooperation? Perhaps the best answer to be given is to refer back to the Christians themselves. Is it not part of the freedom of a Christian to decide how to promote neighborly love and by this peace?

## 2.3 From the Personal Perspective

The functionally differentiated coexistence of Church and state that was established in Christianity (Concordat of Worms, Luther) contributes to the negative peace, because on one hand, it bans any kind of religious war. Solely the state has the monopoly on the use of force, never the Church. On the other hand, this right of the state is restricted to secular affairs. This implies that the state has to accept the freedom of conscience and the freedom of faith. The freedom of religion as a human right is an implication of Christian faith, although it was, regrettably, a long road to establish these rights and, moreover, the Churches stood often in the way and deferred this development.[75] Nevertheless, there are strong and vivid connections between Christian faith and the freedom of conscience, faith, and religion.

As early as the apostle Paul, the Christian faith was combined with the term freedom. “Stand fast therefore in the liberty with which Christ has made us free, and be not entangled again with the yoke of bondage” (Gal. 5:1). Jesus Christ has freed the Christians from religious bondage. Through the Son of God the Christians received adoption as sons and daughters; they have by this attained full age and are neither servants nor (little) children any more that need a prescription by the law on how to behave and to live (Gal. 3:23–26).[76] This liberty of the religious adults implies that they need not observe religious or cosmic rituals (Gal. 4:1–11).[77] Through this, the peace with God modifies the standing of the Christians; their freedom as adults from religious paternalism (negative freedom from) implies the freedom to pray to God without intercession of the Church or religious leaders (positive freedom to).[78]

This freedom to pray to God unmediated became important for Martin Luther. In his essay “The Freedom of a Christian” (1520) he took up Paul’s line of thought and spoke of the Christians metaphorically as kings and priests. They are as kings, because nobody has the power to harm their salvation, and they are as priests, because they are allowed to stand before God as saints do.[79] Even more important became Luther’s creed on the Diet (Reichstag) in Worms

---

**75** Cf. Blickle, Peter, *Von der Leibeigenschaft zu den Menschenrechten. Eine Geschichte der Freiheit in Deutschland*, München: C.H. Beck, 2003; Tönnies, Sibylle, *Die Menschenrechtsidee. Ein abendländisches Exportgut*, Wiesbaden: Verlag für Sozialwissenschaften, 2011.

**76** **Gal. 3:23–26**: “But before faith came, we were kept under the law, shut up to the faith which should afterwards be revealed. Why the law was our schoolmaster to bring us to Christ, that we might be justified by faith. But after that faith is come, we are no longer under a schoolmaster. For you are all the children of God by faith in Christ Jesus.”

**77** **Gal. 4:1–11**: “Now I say, that the heir, as long as he is a child, differs nothing from a servant, though he be lord of all; but is under tutors and governors until the time appointed of the father. Even so we, when we were children, were in bondage under the elements of the world: But when the fullness of the time was come, God sent forth his Son, made of a woman, made under the law, to redeem them that were under the law, that we might receive the adoption of sons. And because you are sons, God has sent forth the Spirit of his Son into your hearts, crying, Abba, Father. Why you are no more a servant, but a son; and if a son, then an heir of God through Christ. However, then, when you knew not God, you did service to them which by nature are no gods. But now, after that you have known God, or rather are known of God, how turn you again to the weak and beggarly elements, whereunto you desire again to be in bondage? You observe days, and months, and times, and years. I am afraid of you, lest I have bestowed on you labor in vain.”

**78** Cf. Berlin, Isaiah, *Four Essays on Liberty*, Oxford: Oxford University Press, 1969.

**79** Cf. Luther, Martin, *The Freedom of a Christian* (1520), ch. 14–16, LW 31, 353–356 = WA 7, 26–28.

in 1521, where he was asked to denounce his writings. Luther refused to do so referring to his conscience that is bound to the Holy Scriptures.[80] By this he calls for the freedom of conscience as a right that is above political and religious claims. It was not peace but quarrel, a schism between the catholic and the Protestant Churches, and later on even wars were evoked by Luther's insistence.

Nevertheless, the appeal on the freedom of conscience became at the same time an argument in the late scholastics in Spain dealing with the mission of the "Indians" in the new territories (America). One important biblical reference was Luke 14:23: "*compelle intrare*" (compel them to come in). Augustine legitimated with this verse the combatting of heretics to return to the Catholic Church, and in the 16th century theologians took up this account to legitimate the forced Christianization and baptism of the native people in America. In contrast, Bartolomé de Las Casas (1484–1566) and Francisco de Vitoria (1492–1546) argued for a mission with spiritual and not violent means. Following Jesus, the Church should convince not with violence but with persuasion.[81] Since the Indians could understand the preaching and are able to believe in Jesus Christ, they have from nature (as a gift of the creator) human dignity and rights.[82] Consequently, they must not be compelled to Christianity, but can become Christians as a question of their own conscience.

If one's conscience is bound and can be convinced only with spiritual means, neither religion nor politics will have the right to urge the conscience. Therefore, the freedom of conscience and of faith must be accepted by both regiments as an individual human right. Additionally this freedom must not be bound to one religion or one belief-system. Regretfully it took decades and

**80** Luther's creed as it was traded in my free translation: "I cannot revoke all my writings, because they are too different. The first part of them is interpretations of the Bible that are widely accepted in the church. Should I cancel them, I would drag the Word of God through the mire. The second part comprises critique of the church and the Pope that is well founded. It is not the task of the political leaders to ban an inner-church position; with that the emperor would only support the Tyranny of the Pope. So I will not cancel these scripts as well. Finally, there are books, in which I have criticized other Christians. It could probably be that I have adopted the wrong tone, but this had to be demonstrated concretely. Unless I am convinced by the testimony of the Scriptures or by clear reason, I am bound by the Scriptures I have quoted and my conscience is captive to the Word of God. I cannot and will not recant anything, since it is neither safe nor right to go against conscience. Here I stand. I can do no other. May God help me. Amen."

**81** Cf. Gillner, Matthias, *Bartolomé de Las Casas und die Eroberung des indianischen Kontinents. Das friedensethische Profil eines weltgeschichtlichen Umbruchs aus der Perspektive eines Anwalts der Unterdrückten*, Stuttgart: Kohlhammer, 1997, 240–243.

**82** Cf. Justenhoven, Heinz-Gerhard, *Francisco de Vitoria zu Krieg und Frieden*, Köln: J.P. Bachem, 1991, 57–61.

awful wars of religion in Europe to gain this insight.[83] Furthermore, considering that faith includes not only a system of convictions, but also religious practices and assemblies of the believers, we ought to speak about the freedom of religion. This term comprises the human location of this freedom (conscience), its content (faith) and its social manifestation (religion). Whereas freedom of religion forbids neither the religions nor the believers to preach the gospel and to proselytize, yet, it limits these efforts to peaceful means – according to the insight that only the state is allowed to use violent force, and that the state at the same time is limited to the *temporalia*. From the personal perspective, the freedom of religion is now added as another limitation for the government and the Church.

With reference to this human right of religious freedom the previous controversy on the acceptance of violent means by Christians can be mitigated. A Christian can decide by him- or herself and for him- or herself whether he or she may become a soldier or a police-officer. The Churches and other Christians can, of course, explain their conviction in this quarrel and try to persuade or convince him or her. Yet, his or her conscience and his or her decision must not be overridden. The same applies to other institutions in society; nobody must be coerced to become a butcher or to work in an abortion clinic against his conscience. Within a government it is, in fact, allowed to ask all civilians for a military conscription or even to install a general conscription. If some people, however, have a moral objection to military service, the state must accept this decision, but may order a substitute service instead. Even the rulers are limited by rules, more precisely by the rule to accept the freedom of conscience. This limitation opens the opportunity for humans, to lead their life as they want to, just following their convictions. In this way, the freedom of religion promotes peace. Furthermore, tolerance is required from the government, the Churches and all citizens as well. They must tolerate the religious convictions of other citizens that they do not share.

However, the combination of tolerance and freedom of religion evokes another challenge for Christianity: Are there any restrictions for tolerance? The classical answer concerning the government is that the freedom of every civilian is limited by the freedom of the others (Immanuel Kant).[84] Thus, religious convictions must be tolerated as long as they do not bother fellow citizens. Whereas thoughts are free, religious speeches and rituals may be limited, since

**83** Cf. Leonhardt, Rochus, *Religion und Politik im Christentum. Vergangenheit und Gegenwart eines spannungsreichen Verhältnisses*, Baden-Baden: Nomos, 2017, 209–44.

**84** Cf. Kant, Immanuel, *Metaphysik der Sitten. Erster Teil. Metaphysische Anfangsgründe der Rechtslehre*, [Königsberg 1797], Einleitung § B: „Das Recht ist also der Inbegriff der Bedingungen, unter denen die Willkür des einen mit der Willkür des anderen nach einem allgemeinen Gesetze der Freiheit zusammen vereinigt werden kann“, Bernd Ludwig (ed.), Hamburg: Meiner, 1986, 38.

they can damage other human rights, and in this case a compromise is needed. Such an occasion was the dispute about circumcision in Germany in 2012, in which the religious duty of the Jews and the Muslims as well as healthcare (as a human right) had to be balanced and written into law. Today the circumcision of male children is allowed in Germany but the circumciser must be medically educated. Living together peacefully requires every religious citizen to accept compromises not only concerning his or her faith, but also in dealing with the social impacts of any religion.

Social compatibility, then, becomes the benchmark to limit religious speeches and rituals. However, how should Christians and the Churches deal with incompatible beliefs? Since citizens are free to form their religious beliefs, the government is originally not involved in answering this question; it has to treat all confessions of faith equally. Christians and Churches by contrast are challenged when the required tolerance contradicts their beliefs. For example it is moot whether children or only adults should be baptized. Both creeds cannot be true at the same time, but both refer to the conscience of believers. Can a conscience make a mistake? If so, how should the Church or the state deal with an erring conscience? These questions were scrutinized by Thomas Aquinas.[85] To answer them, he distinguishes two perspectives. In the theoretical perspective he insists that decisions can be objectively right but also wrong. Humans have by nature moral insights, so they can conceive in their conscience (*συντήρησις* / syntheresis) principally what is right or wrong; for example anybody can recognize that it is wrong to kill an innocent. The individual conscience (*συνείδησις* / syneidesis) however is not infallible. Thomas stresses, contrary to the scholastic tradition, that the individual conscience is not the immediate voice of God, but a natural ability of humans to understand, to pick up or to repudiate God's guidelines. To pick up the example: It is possible that one errs and is convinced it is just to kill a completely innocent person; maybe because he finds him guilty, or maybe he has not accessed the objective insight. The individual conscience is, so to speak, the judicial power of humans that refers to natural insights (common to all humans) and transforms them individually into moral decisions. For this reason, the question of conscience must be accepted in a practical perspective; therefore it would be a sin not to follow one's conscience.[86] The conscience's decision is obligatory for any human, because it is part of his or her identity. Thus,

**85** Cf. for the following Schockenhoff, Eberhard, *Wie gewiss ist das Gewissen? Eine ethische Orientierung*, Freiburg/Breisgau: Herder 2003.

**86** Cf. Aquinas, Thomas, *Quaestiones quodlibetales III*, questio 12, articulus 2c: "Et ideo dicendum est quod omnis conscientia, sive recta, sive erronea, (. . .) est obligatoria; ita quod qui contra conscientiam facit, peccat."

humans must follow their consciences, but they must also carry the social and personal consequences. This argumentation is convincing. Thomas states that the conscience is the individual authority but not the theological benchmark for religious beliefs. The conscience can err and follow a wrong advisor, but it is nevertheless the interior court of every human and part of his or her identity. Thus, the freedom of religion, conscience and faith has to be a human right, as it belongs to his self-determination. Its acceptance will support peace, because it allows humans to follow their deepest convictions.

The challenge for Christians and Churches, however, is still pending. To accept the freedom of religion as limitation for social arrangements and legal regulations does not imply to agree to contradictory beliefs. Given the possibility that these can be wrong, as Thomas has argued, the query even increases: One cannot tolerate what is not true. The term truth implies intolerance. If an assertion is true, contradictory statements are excluded from truth (principle of excluded contradiction). Moreover, every assertion claims uncircumventably to be true. This statement does not imply that every belief-system is true, albeit it implies that every creed affirms itself to be true. Unfortunately, this inevitability of truth leads to an ongoing struggle of religions, since they proclaim assertions. Alternatively, tolerating all of them would imply sidelining their aspirations, as well. Yet, both attitudes do not foster peace but enmity. On the one hand, a struggle between religious claims can easily extend and refer not only to their assertions but also to the believers, who are convinced of their truth. Alas, Church history contains many examples showing that the repudiation of heretic insights was accompanied by the condemnation and punishment of the heretics who were simply bound to their consciences. On the other hand, the toleration of contradictions will only be possible, if none of them matter – and that would be an imposition for the faithful, because it does not take their conscience seriously. Yet to declare a creed to be an adiaphora (incidental), will evoke enmity as well. Since both attitudes can turn perilous, the role of the secular government became more and more important: The state's monopoly on the use of force can guarantee the validity and enforceability of human rights, including the freedom of religion and its limitations – more about this in the next section.

In philosophy, assertions must be verified or falsified to validate their truth-claims. How can one verify or falsify theological assertions that are grounded in a conviction and are anchored in the conscience? This challenge leads us to the theological benchmark and thereby to the religious perspective.

## 2.4 From the Religious Perspective

Being justified by faith, Christians have peace with God through Jesus the Christ (Rom. 5:1). Peace with God is a consequence of salvation and thereby of the forgiveness of sins. In order to interpret this keynote of the New Testament, an accurate understanding of sin is important. Sin is not fixed to the perpetrator (as guilt) but describes the connection between God and humans. More precisely, it labels this relationship as disturbed or even destroyed because of the human's attitude and behavior. Today many Christians agree with Martin Luther that the sin of mankind lies in the tendency to make oneself a god. Humans do not accept God being God; they prefer themselves to be God.[87] They want to rule the world; they want to judge other people. Luther has unfolded this insight in his protest against the letters of indulgence in 1517. Whoever buys or sells a letter of indulgence makes God a broker and thereby does not accept God being God. In 1520, Luther widens this argument:[88] Whoever insists on his own achievements before God, be they financial (indulgence) or moral (good deeds), wants to make a deal with God and by this does not trust in God's promise that all sins will be forgiven. However faith is simply this: trust in God. Following Luther, a Christian will not insure him- or herself with good deeds, because in this case he or she would not trust in God but rely on him- or herself. Precisely this self-centered disposition is what Luther calls the basic sin. This sinner does not accept God's promise but declares the standards of the Final Judgement on his own. By this he downgrades God to a broker and now wants to make a deal with God according to his own benchmarks.

Luther's understanding of sin has an important implication for the freedom of religion and its challenge for Christianity: To live in peace with God means not to insist on being the judge. Whoever trusts in God and his salvation must neither try to deal with him nor replace God's final judgements with his or her own assessments. Repentance implies climbing down from the judgement seat and faith implies trusting God sitting on the judgement seat.[89] Paul has already expressed a similar line of thought in his letter to the Romans:

**87** Cf. Luther, Martin, *Disputation against Scholastic Theology*, 1517: "Man is by nature unable to want God to be God. Indeed, he himself wants to be God, and does not want God to be God." (LW 31, 10 = WA 1, 225, 1–2: „Der Mensch kann von Natur aus nicht wollen, dass Gott Gott ist; er möchte vielmehr, dass er Gott und Gott nicht Gott ist").

**88** Cf. Luther, Martin, *The Freedom of a Christian* (1520), ch. 11, LW 31, 350 = WA 7, 25.

**89** Cf. Ricoeur, Paul, *Geschichte und Wahrheit*, München: Paul List, 1974, 35.

> Recompense to no man evil for evil. Provide things honest in the sight of all men. If it be possible, as much as lies in you, live peaceably with all men. Dearly beloved, avenge not yourselves, but rather give place to wrath: for it is written, Vengeance is mine; I will repay, said the Lord. (Rom. 12:17–19)

Christians should behave as humans and not as if they were God. To accept God being God and oneself being human implies an undisturbed relationship between God and man, in which both have their own profile and none is overruled by the other. This attitude is accompanied by a self-relativization of Christian faith that has two facets. Firstly, God is the judge, neither the Christians nor the Church are. Final judgements are therefore not due to either of them. They shall proclaim the gospel including its warnings of false deities, but they must not make final decisions nor condemn humans because of their religion. Secondly, God is the merciful father who forgives sins. So Christians are depending on the grace of God as well, since they are also sinners. This insight rejects religious hubris and evokes humility instead.

Thus, the freedom of religion as a human right can be supported by Christians not only because they benefit from it (as a religion), but furthermore because all humans are created by God and depend on his grace. The Last Judgement will verify or falsify the Christian faith as well as all other religious beliefs. God will judge all humans and thereby reveal the right and wrong religions. This eschatological caveat does not revoke the truth-claims of Christian creeds, yet it insists to differ between a religious assertion and a human believer. Passing eternal judgement on humans is solely up to God; to accept, to criticize, or to oppose religious assertions falls into the responsibility of Churches and is limited to nonviolent means (*sine vi humana, sed verbo*). Finally, all of these measures promote peace, because they involve religions into the public discourses and prevent by this that a religion absolutizes her own insights.

# 3 Current Tasks Regarding Peace

For Christian faith peace with God as a consequence of our justification is the foundation for peace on earth. This peace with God was realized by Jesus the Christ, in whom God reconciled the world with himself, not imputing their trespasses to them (2Cor. 5:19). From now on, God and Christians will cooperate to realize peace on earth. This earthly peace is two-fold; negative peace denotes the absence of war and violence, positive peace comprises justice and righteousness among the people. In order to support the process of peace, Christian faith accepts the state's monopoly of the legitimate use of force and does not allow the religions to acquire this position. Likewise, the Christian tradition limits this use of force, for example with the just war theory. Furthermore, the Church is called to promote this peace process by proselytizing through peaceful means and engaging in society diaconally. Finally, all must accept freedom of religion as a human right.

It is not sufficient, however, solely to declare human rights; they must also be instilled and guaranteed. Since human rights belong to the *temporalia*, the secular government became more and more important as the authority to enforce these rights. During the religious wars in Europe in the 16th and 17th centuries, the state became the principal guarantor for negative peace. Especially Thomas Hobbes (1588–1679) put all his hope in the absolute ruler as a mortal God (Leviathan). This ruler shall not only have the monopoly on the use of force, he shall furthermore also decide about the religious membership of his state and its citizens, because only this plenitude of power can, following Hobbes, prevent a "*bellum omnia contra omnes*" (a civil war all against all).[90] In order to limit or, if possible, to avoid religious wars, Hobbes strengthened the secular government and rejected religion. Other political philosophers followed him and declared that the *temporalia* should be treated as if there were no God ("*etsi Deus non daretur*"),[91] instead referring to the rights and the laws of the state. Many Christians, therefore, emigrated from Europe to America, because their religious freedom was not guaranteed by the state's laws.

Thus, the rule of law was a central means to achieve negative peace on earth, and it was bound to the state and his monopoly on the use of force. With this modification the cogency of the just war theory was affected, because it was now the state deciding whether to wage a war or not. The states developed into national states and their sovereignty was neither limited by nor bound to the Church or the religion. Each national state as a sovereign entity then has the right to conduct warfare. The strong points of this development were the containment of religious violence and the guarantee of the rule of law. The weak point was the absolute sovereignty of the national state that was neither bound to international institutions (as it will become in the 20th century through the United Nations) nor restricted by responsibility to the citizens (as it will become in the 21th century: the Responsibility to Protect). In summary, the state's sovereignty

**90** Cf. Schotte, Dietrich, *Die Entmachtung Gottes durch den Leviathan. Thomas Hobbes über Religion*, Stuttgart: frommann-holzboog, 2013; Münkler, Herfried, *Thomas Hobbes. Eine Einführung*, Frankfurt/Main: Campus, 2014.

**91** CF. Grotius, Hugo, *De Jure Belli ac Pacis. Libri tres* (1625), ed. By Walter Schätzel, Tübingen: Mohr-Siebeck, 1950, 33; Luther, Martin, *Der 127. Psalm ausgelegt an die Christen zu Riga von 1524* (= WA 15, 373, 3).

ensures the rule of law and fosters negative peace, but it is also dangerous because of the plenitude of power.

Accordingly, peace on earth is still long in coming. Although Steven Pinker has suggested that violence was steadily declining during human history,[92] it does not feel as if we are living in peaceful times. The Second World War and the constant threat of atomic warfare especially challenge Christian peace-ethics: Are atomic weapons, including the ability to destroy the entire planet, still to be categorized as a "sword," and by this justified as just military means? Is the policy of deterrence aiming at negative peace? How should the so called new wars, mostly civil-wars, riots, and acts of terrorism be classified? Neither Christians nor the Churches are able to answer these questions extensively by themselves. Albeit, they should provide a framework or suggestions, based on peace with God and oriented towards contemporary challenges. Consequently, this chapter will deal with proposals of the Christian faith concerning peace on earth in the 21th century.

## 3.1 From the Political Perspective

"War is contrary to the will of God"[93]– with this ethical imperative the World Council of Churches (WCC) in 1948 phrased an insight that the Peace Churches had already expressed, and that had become evident facing the world wars and their atrocities. Meanwhile, both of the major Churches in Germany followed this concept and have performed a paradigm shift, replacing the traditional term "just war" by the concept of "just peace."[94] The Catholic German Bishops Conference as well as the Protestant Council of EKD have recently each published

**92** Cf. Pinker, Steven, *The Better Angels of Our Nature. Why Violence Has Declined*, New York: The Viking Press, 2011.

**93** Cf. World Council of Churches, *Just Peace Companion*, 2nd edition 2012, 15; in German: Bericht der Vierten Sektion der Gründungs-Vollversammlung des Ökumenischen Rates der Kirchen in Amsterdam 1948; in: Kirche und Frieden. EKD Texte 3, Hannover (Kirchenkanzlei der EKD) 1982 155–162, 156: „Kriege sollen nach Gottes Willen nicht sein". Cf. Garstecki, Joachim, "Ist noch drin, was draufsteht? Ökumenische Friedensethik und kirchliche Friedensarbeit im Spannungsfeld zwischen ziviler Konfliktbearbeitung, militärischem Interventionismus und öffentlicher Kriegsgewöhnung. Eine Problemanzeige," in: Friedemann Stengel/ Jörg Ulrich (eds), *Kirche und Krieg. Ambivalenzen in der Theologie*, 213–231, Leipzig: EVA, 2015.

**94** Cf. Werkner, Ines-Jacqueline, *Gerechter Friede. Das fortwährende Dilemma militärischer Gewalt*, Bielefeld: transcript, 2018.

a memorandum with “just peace” in the title.[95] This current conception of just peace lies in the line of sight of the assertion from 1948.[96] In further publications, the WCC has unfolded this assertion and has also taken up the term “just peace.”[97] The notion of just peace rejects the traditional nexus in Christianity between war and justice as is found in the just war theory. Conflicts and quarrels are from biblical viewpoint indeed a fact that can be described and should be evaluated ethically afterwards. Thus, not the conflicts but the means to solve them should be moral and can be criticized from an ethical point of view as well. However, this applies to conflicts, and not to wars, following the WCC. War is not an occasionally arising incident like any other conflict that can be evaluated ethically as just or unjust. Correspondingly, war is not merely the continuation of policy by other means.[98] War rather indicates a failure of politics, as the main assignment of politics is to take responsibility for peace, justice, order, and law.

Whereas the just war tradition accepted war as a political means and limited it with ethical criteria, the just peace conception repudiates the idea that war can be a legitimate act.[99] Only peace can be referred to as just, never war. At best, a war might be the lesser evil and therefore risked – but it is still an evil. At first this could sound like a mere controversy on words. Yet there is an underlying shift of paradigm: Political considerations shall be designed to start not from war but from peace. To achieve peace means to prepare peace – and

---

**95** Cf. Die deutschen Bischöfe, *Gerechter Friede, Bonn 2000; Aus Gottes Frieden leben – für gerechten Frieden sorgen. Eine Denkschrift des Rates der Evangelischen Kirche in Deutschland* (EKD), Gütersloh: Gütersloher Verlagshaus, 2007.

**96** For the Catholic peace-ethics the papal encyclical “pacem in terries,” published in 1963 during Vaticanum II was a milestone; cf. Justenhoven, Heinz-Gerhard/O'Conell, Mary Ellen (eds), *Peace Through Law. Reflections on Pacem in Terris from Philosophy, Law, Theology, and Political Science*, Baden-Baden: Nomos, 2016. For the Protestant traditions Dietrich Bonhoeffer was quite influential; cf. von Lüpke, Johannes, “Frieden im Kampf um Gerechtigkeit und Wahrheit. Dietrich Bonhoeffers Friedensethik,” in: Volker Stümke/ Matthias Gillner (eds), *Friedensethik im 20. Jahrhundert*, 13–28, Stuttgart: Kohlhammer, 2011.

**97** Cf. World Council of Churches, *Just Peace Companion*, 2nd edition 2012. Cf. Raiser, Konrad, “Eine Ethik rechtserhaltender Gewalt im ökumenischen Diskurs. Zwischen gerechtem Krieg und Pazifismus,” in: Ines-Jacqueline Werkner/ Torsten Meireis (eds), *Rechtserhaltende Gewalt – eine ethische Verortung. Fragen zur Gewalt Band 2*, 95–115, Wiesbaden: Springer, 2018.

**98** Cf. von Clausewitz, Carl, *Vom Kriege* [1832], Neuausgabe, München: Ullstein, $^{3}$2002, Erstes Buch, 1. Kapitel, Abschnitt 24, 44: „Krieg ist die bloße Fortsetzung der Politik mit anderen Mitteln“.

**99** Cf. Haspel, Michael, “Die „Theorie des gerechten Friedens“ als normative Theorie internationaler Beziehungen? Möglichkeiten und Grenzen,” in: Jean-Daniel Strub/Stefan Grotefeld (eds), *Der gerechte Friede zwischen Pazifismus und gerechtem Krieg. Paradigmen der Friedensethik im Diskurs*, 209–225, Stuttgart: Kohlhammer, 2007.

not war: *si vis pacem, para pacem* – and not, as the old saying goes: *para bellum* (prepare the war).[100] In this context of "*para pacem*" (prepare peace), the international law, just policing, good governance, and international networking become very important. Hence, just peace corresponds to legal pacifism. Both refer the prohibition of force according to Art. 2 Par. 3f of the United Nation Charter:[101] conflicts and disputes ought to be solved with peaceful means; an international legal framework is a very important means.

*Para pacem* – this imperative already implies the two main insights from the just peace paradigm: Firstly peace is a process of preparing peace and rejecting violence; it is neither a constant factor nor an unchangeable ideal. Secondly this peace process is complex and therefore depends on networking (in politics: comprehensive approach). This multifacetedness can be illustrated with the aims that the WCC has stated in 2012:

- for peace in the community – so that all may live free from fear (Mic. 4:4),
- for peace with the earth – so that life is sustained,
- for peace in the marketplace – so that all may live with dignity,
- for peace among the people – so that human life is protected.[102]

These aims are linked to a peace-process in two perspectives. On the one hand the concept itself is evolving; analyzing "*para pacem*" implies discovering connections and achievements that can hamper or foster peace. Violence and force have likewise many facets that are partially interconnected; thus, new challenges can be detected. On the other hand these aims are signposts that will lead a certain way to promote peace. Yet, since people as well as communities live and change, these aims will never be realized in full. Thus, promoting, preserving, and renewing peace is a perpetual endeavor.[103]

**100** Cf. Senghaas, Dieter, "Frieden als Zivilisierungsprojekt," 14. An earlier form of this concise formulation can be found in the 19th century liberalism: „si vis pacem, para libertatem et iustitiam" – cf. Czempiel, Ernst-Otto, *Friedensstrategien. Eine systematische Darstellung außenpolitischer Theorien von Machiavelli bis Madariaga*, Opladen: Verlag für Sozialwissenschaften, 21998, 165f.

**101** Cf. UNC (1945) art. 2 par. 3f: "All Members shall settle their international disputes by peaceful means in such a manner that international peace and security, and justice, are not endangered. All Members shall refrain in their international relations from the threat or use of force against the territorial integrity or political independence of any state, or in any other manner inconsistent with the Purposes of the United Nations."

**102** Cf. World Council of Churches, *Just Peace Companion*, 2nd edition 2012, 9–13.

**103** Cf. *Aus Gottes Frieden*, 2007, 11. – In the following passages I will outline the main insights from the current church papers dealing with just peace. As there are many consonances I will not refer to certain papers nor quote them directly.

The peace process is therefore connected with the conflict process. In both cases one can differ between prevention, intervention, and post-conflict peace building. Prevention does not imply prohibiting conflicts; quite the opposite, they can be helpful for individuals and for society to define their way of life or to set their priorities for themselves. Conflict should be prevented from turning violent. Education for peace, just conditions in society, equivalent opportunities in the market, and healthy environment will help reaching this goal. The Churches currently stress prevention as most important and they list many measures to foster prevention of conflict such as supporting the rule of law. A constitutional state with the separation of powers might be the best protection against new wars, especially in failing states. Yet the implications for the international system are debated:[104] When peace can be supported through law, do we need a world-state to establish and foster this law, although it might become tyrannical? Or is it better to merely install an international court, although it will be basically powerless, since the national states will not accept to be overruled by an alien organization? In any case, international organizations should be strengthened and human rights must be accepted worldwide. From the military perspective, the delivery of arms must be controlled, disarmament (not only of nuclear weapons) must make progress, and the privatization of violence must be stopped.

Before we turn to the issue of intervention into a violent conflict, some post-conflict remedies must be discussed. Since the rule of law is one of the most important aims, the stability of the state must be promoted to build up peace. At the same time, atrocities and injustice must be dredged up, and this is a painful and stressful process, because there are two aims that are both important but cannot be reached together. On the one hand, people want to know the truth, for example they want to know who violated the rights of their relatives. On the other hand, they want justice and atonement. Thus, the perpetrators are afraid to reveal the truth because of the legal consequences. Accordingly, the jurisdiction decrees that no one can be forced to incriminate oneself. A truth commission and the possibility of forgiveness should therefore be combined.[105] And it should be placed in the local governments so that victims and perpetrators can narrate their experiences and together find a solution to reconciliation. However, this procedure must be limited; otherwise the state's monopoly on legislation and jurisdiction will be weakened.

**104** Cf. Justenhoven/O'Conell, *Peace Through Law*, 2016.

**105** Cf. Gobodo-Madikizela, Pumla, *A Human Being Died That Night. A South African Woman Confronts the Legacy of Apartheid*, London: Granta Books, 2006.

The most defying point in the conflict process is reached when conflicts turn violent. In this case the peace process shall concentrate on protection and mediation. Whereas Churches and faith-based actors have been especially helpful in mediating (see below in the following subsection), politics is mainly responsible for the protection of the people. More precisely, two aspects are debated in the current Church papers. On the one hand the concept of "responsibility to protect" (R2P), developed by an international commission sponsored by the Canadian Government (ICISS) in 2001, caused a shift in international law: Not the interests of the (national) state, but the concerns of the people are crucial for the legacy of the government.[106] The government must protect its people. "Each individual State has the responsibility to protect its populations from genocide, war crimes, ethnic cleansing and crimes against humanity."[107] If it cannot or will not fulfill this duty, the international community may intervene with civil and military means up to a humanitarian intervention as the last resort (ultima ratio). This political commitment picks up the vote of the Church to protect the vulnerable, especially women and children, albeit with the possibility to use military means and thereby stuck in the cycle of violence.[108] Therefore, Christians are split in their attitude towards R2P.

On the other hand, the debates carry on between the two major peace traditions relating to the use of military force – as can be seen in the reactions to R2P. With the concept of just peace, the two major Churches in Germany have taken up the insights from the Peace Churches that war cannot be a just means. The Churches, following Jesus Christ, are bound to peace and shall therefore foster civil activities. However, in an armed conflict the vulnerable must be protected, if necessary with police and military means. At this point, the queries start with two questions: Is the state justified in using armed force? And are Christians allowed to support the state, for example as soldiers? These topics are still highly controversial among Christians and in the Churches.[109] The concept of just peace in the major Churches results in a legal pacifism, which implies that the use of force cannot be ruled out in any case, but must be restricted rigorously; whereas

**106** Cf. Evans, Gareth, *The Responsibility to Protect. Ending Mass Atrocity Crimes Once and for All*, Washington: Brookings Institution Press, 2009; Verlage, Christopher, *Responsibility to Protect. Ein neuer Ansatz im Völkerrecht zur Verhinderung von Völkermord, Kriegsverbrechen und Verbrechen gegen die Menschlichkeit*, Tübingen: Mohr-Siebeck, 2009.

**107** Cf. The United Nations General Assembly, *World Summit Outcome Document* 2005, 138.

**108** Cf. Busche, Hubertus/Schubbe, Daniel (eds), *Die Humanitäre Intervention in der ethischen Beurteilung*, Tübingen: Mohr-Siebeck, 2013.

**109** Cf. Enns, Fernando/Weiße, Wolfram (eds), *Gewaltfreiheit und Gewalt in den Religionen. Politische und theologische Herausforderungen*, Münster: Waxmann, 2016.

the Peace Churches cling to an absolute pacifism including the possibility for Christians to suffer as Jesus did. In order to settle this quarrel, the Churches stress again the processual character of peace, implying that they are all on their way to peace, but starting from different points and taking varying routes.

With this reference to the process of peace, the divergences can be explained and tolerated whereas the commonalities are stressed. For example, in 2007 the EKD has modified its position regarding nuclear weapons.[110] Having experienced the atrocities of the atom bombs in the Second World War (Hiroshima, Nagasaki), the Protestant Churches have since condemned the use of nuclear weapons. However, in the "Heidelberger Thesen" (1959), the tolerance of ownership of these weapons, including the threat to use them (deterrence) was accepted as a Christian behavior at that instant ("noch").[111] Nearly fifty years later (2007), the EKD stated that threatening with nuclear weapons can no longer be seen as a means of legitimate self-defense. This enlarged commonality with the Peace Churches created albeit another controversy: How should the government deal with those nuclear weapons that are already there? Is a nuclear disarmament required, certainly step by step, or must they be kept to hinder a new rat-race to earn these weapons, since the knowledge to build such bombs can never be extinguished?

This example illustrates the strong points of speaking of a peace-process: it allows modifications and facilitates tolerance. Nevertheless, there is also a weak point: A process may have different starting points, but it has one end. From this point of view, the different positions can be measured and compared. Such a ranking is dangerous though, because it upgrades the one and downgrades the other. In order to avoid this danger the second insight from "*para pacem*," the complexity, should be spelled out as a practice in discourse.

110 Cf. Stümke, Volker, "Der Streit um die Atombewaffnung im deutschen Protestantismus," in: Volker Stümke/Matthias Gillner (eds), *Friedensethik im 20. Jahrhundert*, 49–69, Stuttgart: Kohlhammer, 2011; Möller, Ulrich, *Im Prozeß des Bekennens. Brennpunkte der kirchlichen Atomwaffendiskussion im deutschen Protestantismus 195–1962*, Neukirchen-Vluyn: Neukirchener Verlagsgesellschaft, 1999.

111 Schockenhoff, Eberhard, *Kein Ende der Gewalt? Friedensethik für eine globalisierte Welt*, Freiburg/Breisgau: Herder, 2018, 379f explains that and how this „noch" modified it's meaning. In the 1950$^{th}$ three interpretations were at hand; firstly a factual reading (as a political compromise), secondly a psychological reading (as a personal acceptance of fear) and thirdly a temporal reading (as a time limit). Forty years later, this temporal meaning dominated and demanded the politics to start with disarmament talks.

## 3.2 From the Social Perspective

What should be done? The list of necessary or at least useful measures to achieve and preserve peace is long. In the state, a good government is needed, whereas corruption and nepotism are dangerous. The rule of law must be established and furthermore the laws should be just. In the market allocation should be performed in a fair way so that no one will have to starve. Children should be educated and should go to school. Man and woman must be treated equally. People must have the ability to participate in the political decision-making process and their human rights must be accepted and protected. Furthermore, they must be enabled and have the chance to lead a good and self-determined life. The environment must not be overcharged so that future generations can also live peacefully and without fear or need on this earth.

The main problems of this list are at hand. For a start, it is very abstract and does not entail concrete steps. Yet it is the process of developing concrete steps, and not the principles, where controversies begin. Furthermore, not all aims can be pursued at the same time with the same emphasis. Thus the problem of how to rank these aims arises. Finally, some of these aims do not suit each other well. Nevertheless, the aims themselves sound convincing. Consequently, the implementation process is decisive. The Churches could declare these aims in a know-it-all manner, by this mixing up their regiment in *spiritualia* and its duties with God and forgetting about the self-relativization they should have learned (as explained in the preceding section). Besides, it is of course not a peaceful behavior to act as guardian for the fellow citizens.

The better decision for the Churches would be to practice discourse. No less a figure than Pope Benedict XVI has advised the Church to engage in discourse, mainly with philosophy.[112] A critical dialogue of the Christian faith with philosophy is supportive for both sides.[113] On the one hand, the philosophy may learn that metaphysics is still a topic area for reason when it is confronted with Christians who can explain what and why they believe. On the other hand, Christians and the Churches may become more wary of turning

**112** Vgl. Benedikt XVI., "Glaube, Vernunft und Universität. Erinnerungen und Reflexionen," in: Christoph Dohmen (ed.), *Die „Regensburger Vorlesung" Papst Benedikts XVI. im Dialog der Wissenschaften*, Regensburg: 15–26, Pustet, 2007, and id., "Ansprache seiner Heiligkeit Papst Benedikt XVI. im Deutschen Bundestag," in: Georg Essen (ed.), *Verfassung ohne Grund? Die Rede des Papstes im Bundestag*, 17–26, Freiburg/Breisgau: Herder, 2012.

**113** Ratzinger, Joseph, "Was die Welt zusammenhält. Vorpolitische moralische Grundlagen eines freiheitlichen Staates," in: Jürgen Habermas/Joseph Ratzinger (eds), *Dialektik der Säkularisierung. Über Vernunft und Religion*, 39–60, Freiburg/Breisgau: Herder, 2005.

into fundamentalists. Fundamentalists tend to lock themselves into their belief-system and regard all others as outsiders and enemies. Yet challenged through the critical questions of philosophy, they cannot hide in their pious shells but must justify themselves and their confessions – or modify them when the philosophical annotations are compelling. This papal approach to connect faith and reason promotes the social engagement of the Churches to achieve and preserve peace as well. Instead of "knowing it all" from a pretended transcendent point of view, Churches can connect with other social institutions. And this connection is not a silent coexistence, but cooperation founded on discourse.

In supporting the peace process the Churches have a particular task not above, but beside the other contributors: According to Jesus Christ, the Church itself shall be a sign of peace and reconciliation or, as the German Bishops more precisely and provocative stated: a sacrament of peace.[114] The Churches shall point with their existence beyond themselves to the reconciliation with and through God. In their preaching and in the liturgy, in their own repentance and in the willingness to forgive, in their engagement for the poor and needy (as advocates for justice) and in political counselling, and last, but not least in their own dependence on God's forgiveness, the Churches proclaim and express peace with God becoming reality on earth. For this reason, the Churches must not use violent means themselves. Hence, Christians shall engage in social ministry for the good of humanity: By doing so they begin to implement the required justice in society; for example, they commit themselves globally to fair trade and locally to emergency relief. Whilst Christians and Churches shall take further steps towards peace, neither may undertake God's task: salvation. Christians and Churches are thereby protected against the risks to overburden their options for action, and from self-deification. This protective limitation will increase their placidity and their peace-ability, because it refers them to their place and tasks.

## 3.3 From the Personal Perspective

Not only the Churches, but also the Christians are called to support the peace process. A brief look to the role of faith-based actors in armed conflict shall illustrate this specific challenge. The political scientist Markus A. Weingardt, mostly engaged in peace and conflict studies, has examined how religions, and especially their followers (he speaks of faith-based actors), have been catalysts to

114 Cf. Die deutschen Bischöfe, *Gerechter Friede*, Bonn, 2000, 89.

the peace-progress in various conflicts worldwide.[115] According to Weingardt, these local actors are very successful for three reasons: Firstly, they are emotionally involved in the conflict, since they love the country and its people. Secondly, they are trustworthy, because their aim to achieve peace is not tarnished by their own economic or political interests; it is rather based on their faith and concentrated on ending the conflict. Thirdly, they are thought to be capable of achieving peace, since they do not belong to one of the parties of the dispute. Altogether, the faith-based actors have a credit of trust. This applies to Christians and Church representatives as well. They must use this credit by engaging in the peace-process.

Furthermore, peace education is an important means, and education is primarily located within the family. The WCC publication from 2012 defines education as "a profoundly spiritual formation of character that happens over a long period of time." This is, according to the WCC, a "holistic process of character formation," and the "everyday life practise" that shall be influenced "from the very beginning" – by parents, Christian teachers as well as the Churches. "That involves introspection of all members of the Church, into how their choices, their actions and their lifestyles do or do not make them servants of peace."[116] Nevertheless, this goes too far. Christian educators tend to completely engulf the child and thereby damage the freedom of religion. Furthermore, this holistic approach tries to do the work of the Holy Spirit. Peace education is indeed important and should be practized not only in word but also in deed and cogent behavior. Yet it must know its own limitations, otherwise peace education would spread itself too thin and would tend to a self-deification.

## 3.4 From the Religious Perspective

Christianity as a religion can foster peace but it can also become evil and support religious wars. Charles Kimball listed five features that indicate "when

115 Cf. Weingardt, Markus, *Religion Macht Frieden. Das Friedenspotential von Religionen in politischen Gewaltkonflikten*, Stuttgart: Kohlhammer, 2007; id., *Was Frieden schafft. Religiöse Friedensarbeit. Akteure, Beispiele, Methoden*, Gütersloh: Gütersloher Verlagshaus, 2014.

116 World Council of Churches, *Just Peace Companion*, 2nd edition 2012, 113–115.

religion becomes evil."[117] Though Kimball has scrutinized the three monotheistic religions, I will concentrate now on Christianity:

1. The first danger of religion is its aspiration to claim absolute truth, since this aspiration may easily lead to intolerance, and intolerance becomes dangerous when not only arguments and truth-claiming statements are attacked, but humans are affected as well. Letting this argument play out, to be intolerant against the content of an argument is not half as dangerous as the intolerance against humans. Christian faith guard against this danger and should be clear in its mind time and time again, that God is the judge and that God will vouch for the eternal truth. Christians proclaim the Triune God, but he alone can and will prove that he is Father, Son and Holy Spirit.
2. The second danger lies in faith itself. Faith bound to God excessively can lead to blind obedience. Such Christians isolate themselves and condemn others, first inside the group, then expand to others in other constellations. To prevent this danger, debates with other religions, with philosophy, humanities, and sciences are vital for Christianity.
3. Kimball calls the third danger the "establishing of the ideal time." Those who are convinced that the end is near will not deal with the secular world and its daily challenges. In this case, Luther's legitimization of the secular government and his call for Christians to support the state's work with their professions can help to cool down religiously overheated minds.
4. The fourth danger is the hubris of the Church and the faith. He or she who is determined to have the right and last word wants to dominate others and is not willing to look for balance and compromise. Hopefully, such a Christian will remember that he or she is not God, but a sinner depending on God's forgiveness just like any other.
5. Finally, the fifth danger in Kimball's list is the identification of religion and nation. One who adores the government and its politicians and who swears absolute loyalty to the temporal rulers is as dangerous as those who want to build a religious state where everyone has to obey the religious leader. Both should consider that, according to Luther, God has ordained two regiments.

Christianity must be aware that biblical and religious sources, and furthermore even the confessions and convictions of the Christian faith, carry peaceful messages, but also some precarious contents that are open to violence. The Christian

---

**117** Cf. Kimball, Charles, *When Religion Becomes Evil*, New York, Harper, 2002; id., *When Religion becomes lethal The Explosive Mix of Politics and Religion in Judaism, Christianity, and Islam*, San Francisco: Jossey-Bass, 2011.

faith is not simply exploited by bellicose politicians or war-lords, it has perilous resources.[118] There are two ways to handle this danger. On the one hand, one can limit the political influence of religion.[119] For Christianity, this implies accepting the self-relativizations (the Church is neither God's proxy nor the temporal regiment) and the duties (to proclaim the gospel with words and without violence). On the other hand, one can start a relecture of the precarious texts and analyze the perilous interpretations.[120] By this one can come into dialogue with radical, (i.e. fundamentalist) Christians, because both parties refer to the same sources.

Furthermore, facing these possible perils, all religions should be interested in initiating a world-wide cooperation concerning peace ethic. Especially, in our times of pluralism and globalization, such attempts of understanding among the religions have become urgent. Hence, Hans Küng started the "Projekt Weltethos" in 1990 with three premises:

- No peace among the nations is possible without peace among the religions.
- No peace among the religions is possible without dialogue between the religions.
- No dialogue between the religions is possible without basic research in one's own religion.[121]

In 1993, a Parliament of the World's Religions signed a declaration based on the Golden Rule: "What you wish done to yourself, do to others" (Matt. 7:12).[122] Thus, the religions have found moral values and principles that all can share, although the derivation and the concrete wording might differ. As a result, these shared moral insights are open for divergent reasonings. This interreligious dialogue can foster understanding, tolerance and acceptance between the religions. Its focus is acquiring peace on earth through ethics and puts the definition of God and the associated truth-claims in the rear. The eschatological caveat that God will be the judge and will bring eternal peace is therefore

---

**118** Cf. Assmann, Jan, *Totale Religion. Ursprünge und Formen puritanischer Verschärfung*, Wien: Picus, 2016.

**119** Cf. Svensson, Isak, *Ending Holy Wars. Religion and Conflict Resolution in Civil Wars*, Queensland: University of Queensland Press, 2012.

**120** Cf. Kippenberg, Hans G., *Gewalt als Gottesdienst. Religionskriege im Zeitalter der Globalisierung*, München: C.H. Beck, 2008.

**121** Cf. Küng, Hans, *Projekt Weltethos*, München: Piper, 1990; id., *Handbuch Weltethos. Eine Vision und ihre Umsetzung*, München: Piper, 2012; Frühbauer, Johannes F., "Das Projekt Weltethos," in: Ines-Jacqueline Werkner/Klaus Ebeling (eds), *Handbuch Friedensethik*, 915–924, Wiesbaden: Springer, 2017.

**122 Matt. 7:12**: "Therefore all things whatever you would that men should do to you, do you even so to them."

steadily valid. And correspondingly, humility is still indispensable in achieving peace on earth.

# Bibliography

Amerise, Marilena, "Monotheism and the Monarchy. The Christian Emperor and the Cult of the Sun in Eusebius of Caesarea," *Jahrbuch für Antike und Christentum* 50 (2007), 72–84.

Assmann, Jan, *Totale Religion. Ursprünge und Formen puritanischer Verschärfung*, Wien: Picus, 2016.

Audisio, Gabriel, *Die Waldenser*, Augsburg: Bechtermünz, 2004.

*Aus Gottes Frieden leben – für gerechten Frieden sorgen. Eine Denkschrift des Rates der Evangelischen Kirche in Deutschland*, Gütersloh: Gütersloher Verlagshaus, 2007.

Beestermöller, Gerhard, *Thomas von Aquin und der gerechte Krieg. Friedensethik im theologischen Kontext der Summa Theologiae*, Köln: J.P. Bachem, 1990.

*Die Bekenntnisschriften der evangelisch-lutherischen Kirche* [BSELK], Göttingen: Vandenhoeck & Ruprecht, $^{8}$1979.

Berlin, Isaiah, *Four Essays on Liberty*, Oxford: Oxford University Press, 1969.

Biggar, Nigel, *In Defence of War*, Oxford: Oxford University Press, 2013.

Blickle, Peter, *Von der Leibeigenschaft zu den Menschenrechten. Eine Geschichte der Freiheit in Deutschland*, München: C.H. Beck, 2003.

Busche, Hubertus/ Schubbe, Daniel (eds), *Die Humanitäre Intervention in der ethischen Beurteilung*, Tübingen: Mohr-Siebeck, 2013.

von Clausewitz, Carl, *Vom Kriege* [1832], Neuausgabe München, Ullstein, $^{3}$2002.

Czempiel, Ernst-Otto, *Friedensstrategien. Eine systematische Darstellung außenpolitischer Theorien von Machiavelli bis Madariaga*, Opladen: Verlag für Sozialwissenschaften, $^{2}$1998.

van Dam, Raymond, *The Roman Revolution of Constantine*, Cambridge: Cambridge University Press, 2007.

Die deutschen Bischöfe, *Gerechter Friede*, Bonn: Sekretariat der Deutschen Bischofskonferenz, 2000.

Enns, Fernando, *Friedenskirche in der Ökumene. Mennonitische Wurzeln einer Ethik der Gewaltfreiheit*, Göttingen: Vandenhoeck & Ruprecht, 2003.

Dohmen, Christoph (ed.), *Die „Regensburger Vorlesung" Papst Benedikts XVI. im Dialog der Wissenschaften*, Regensburg: Pustet, 2007.

Enns, Fernando/Weiße Wolfram (eds), *Gewaltfreiheit und Gewalt in den Religionen. Politische und theologische Herausforderungen*, Münster: Waxmann, 2016.

Essen, Georg (ed.), *Verfassung ohne Grund? Die Rede des Papstes im Bundestag*, Freiburg/ Breisgau: Herder, 2012.

Evans, Gareth, *The Responsibility to Protect. Ending Mass Atrocity Crimes Once and for All*, Washington: Brookings Institution Press, 2009.

Flasch, Kurt, *Augustin. Einführung in sein Denken*, Stuttgart: Reclam, $^{2}$1994.

Flebbe, Jochen/Hasselhoff Görge K. (eds), *Ich bin nicht gekommen, Frieden zu bringen, sondern das Schwert. Aspekte des Verhältnisses von Religion und Gewalt*, Göttingen: Vandenhoeck & Ruprecht, 2017.

Forderer, Tanja, "Frieden in den neutestamentlichen Schriften," in: Elisabeth Gräb-Schmidt /Julian Zeyher-Quattlender (eds), *Friedensethik und Theologie. Systematische*

*Erschließung eines Fachgebiets aus der Perspektive von Philosophie und christlicher Theologie*, 117–36, Baden-Baden: Nomos, 2018.

Frowe, Helen, *The Ethics of War and Peace. An Introduction*, London: Routledge, ²2016.

Gillner, Matthias, *Bartolomé de Las Casas und die Eroberung des indianischen Kontinents. Das friedensethische Profil eines weltgeschichtlichen Umbruchs aus der Perspektive eines Anwalts der Unterdrückten*, Stuttgart: Kohlhammer, 1997.

Gillner, Matthias, "Bartolomé de Las Casas und die Menschenrechte," *Jahrbuch für christliche Sozialwissenschaften* 39 (1998), 143–60.

Girardet, Klaus Martin, *Die konstantinische Wende. Voraussetzungen und geistige Grundlagen der Religionspolitik Konstantins des Großen*, Darmstadt: Wissenschaftliche Buchgesellschaft, 2006.

Gobodo-Madikizela, Pumla, *A Human Being Died That Night. A South African Woman Confronts the Legacy of Apartheid*, London: Granta Books, 2006.

Grotius, Hugo, *De Jure Belli ac Pacis. Libri tres* (1625), ed. by Walter Schätzel, Tübingen: Mohr-Siebeck, 1950.

Habermas, Jürgen/ Ratzinger, Joseph (eds), *Dialektik der Säkularisierung. Über Vernunft und Religion*, Freiburg/Breisgau: Herder, 2005.

Hammann, Gottfried, *Die Geschichte der christlichen Diakonie. Praktizierte Nächstenliebe von der Antike bis zur Reformationszeit*, Göttingen: Vandenhoeck & Ruprecht, 2003.

Hengel, Martin, *Die Zeloten. Untersuchungen zur jüdischen Freiheitsbewegung in der Zeit von Herodes I. bis 70 n. Chr*, Leiden/Köln: E.J. Brill, 1976.

Henke, Manfred, *Wir haben nicht einen Bettler unter uns. Studien zur Sozialgeschichte der frühen Quäkerbewegung*, Berlin, be.bra wissenschaft verlag, 2015.

Herrmann, Volker/Horstmann, Martin (eds), *Studienbuch Diakonik Band 1. Biblische, historische und theologische Zugänge zur Diakonie*, Neukirchen-Vluyn: Neukirchener Verlagsgesellschaft, 2006.

Hofheinz, Marco/van Oorschot, Frederike (eds), *Christlich-theologischer Pazifismus im 20. Jahrhundert*, Baden-Baden: Nomos, 2016.

Jäger, Sarah/ Werkner, Ines-Jacqueline (eds), *Gewalt in der Bibel und in kirchlichen Traditionen. Fragen zur Gewalt Band 1*, Wiesbaden: Springer, 2018.

Justenhoven, Heinz-Gerhard, *Francisco de Vitoria zu Krieg und Frieden*, Köln: J.P. Bachem, 1991.

Justenhoven, Heinz-Gerhard/Barbieri Jr., William A., *From Just War to Modern Peace Ethics*, Berlin/Boston: de Gruyter, 2012.

Justenhoven, Heinz-Gerhard/O'Conell, Mary Ellen (eds), *Peace Through Law. Reflections on Pacem in Terris from Philosophy, Law, Theology, and Political Science*, Baden-Baden: Nomos, 2016.

Kant, Immanuel, *Metaphysik der Sitten. Erster Teil. Metaphysische Anfangsgründe der Rechtslehre* [Königsberg 1797], ed. Bernd Ludwig, Hamburg: Meiner, 1986.

Kimball, Charles, *When Religion Becomes Evil*, New York: Harper, 2002.

Kimball, Charles, *When Religion Becomes Lethal. The Explosive Mix of Politics and Religion in Judaism, Christianity, and Islam*, San Francisco: Jossey-Bass, 2011.

Kippenberg, Hans G., *Gewalt als Gottesdienst. Religionskriege im Zeitalter der Globalisierung*, München: C.H. Beck, 2008.

*Kirche und Frieden. EKD Texte 3*, Hannover: Kirchenkanzlei der EKD, 1982.

Knight, George R., *Anticipating the Advent. A Brief History of Seventh-day Adventists*, Nampa: Pacific Pr Pub Assn, 1993.

Küng, Hans, *Projekt Weltethos*, München: Piper, 1990.
Küng, Hans, *Handbuch Weltethos. Eine Vision und ihre Umsetzung*, München: Piper, 2012.
Lambert, Malcolm: *Geschichte der Katharer*, Darmstadt: Wissenschaftliche Buchgesellschaft, 2001.
Lederhilger, Severin J. (ed.), *Gewalt im Namen Gottes. Die Verantwortung der Religionen für Krieg und Frieden*, Frankfurt/Main: Peter Lang, 2015.
Leonhardt, Rochus/von Scheliha, Arnulf (eds), *Hier stehe ich, ich kann nicht anders! Zu Martin Luthers Staatsverständnis*, Baden-Baden: Nomos, 2015.
Leonhardt, Rochus, *Religion und Politik im Christentum. Vergangenheit und Gegenwart eines spannungsreichen Verhältnisses*, Baden-Baden: Nomos, 2017.
Lichdi, Diether G., *Die Mennoniten in Geschichte und Gegenwart. Von der Täuferbewegung zur weltweiten Freikirche*, Weisenheim: Agape, $^{2}$2004.
Luther, Martin, *Werke. Kritische Gesamtausgabe*, Weimar: Böhlau, since 1883 [WA].
Luther, Martin, *American Edition of Luther's Works*, Jaroslav Pelikan/Helmut T. Lehmann (ed.), St. Louis & Philadelphia 1955–1986 [LW].
Markschies, Christoph, *Das antike Christentum. Frömmigkeit, Lebensformen, Institutionen*, München: C.H. Beck, 2016.
McGuckin, John Anthony, *The Orthodox Church. An Introduction to Its History, Doctrine, and Spiritual Culture*, Oxford: John Wiley and Sons, 2010.
McMahan, Jeff, *Killing in War*, Oxford: Oxford University Press, 2009.
Möller, Ulrich, *Im Prozeß des Bekennens. Brennpunkte der kirchlichen Atomwaffendiskussion im deutschen Protestantismus 1957–1962*, Neukirchen-Vluyn: Neukirchener Verlagsgesellschaft, 1999.
Münkler, Herfried, *Thomas Hobbes. Eine Einführung*, Frankfurt/Main (Campus) 2014.
Oeldemann, Johannes, *Die Kirchen des christlichen Ostens. Orthodoxe, orientalische und mit Rom unierte Kirchen*, Düsseldorf: Topos, 2016.
Otto, Eckhart, *Krieg und Frieden in der Hebräischen Bibel und im Alten Orient. Aspekte für eine Friedensordnung in der Moderne*, Stuttgart: Kohlhammer, 1999.
Packull, Werner O., *Hutterite Beginnings. Communitarian Experiments during the Reformation*, London: Johns Hopkins University Press, 1995.
Pinker, Steven, *The Better Angels of Our Nature. Why Violence Has Declined*, New York: The Viking Press, 2011.
Ricoeur, Paul, *Geschichte und Wahrheit*, München: Paul List, 1974.
Rodin, David/Shue Henry (eds), *Just and Unjust Warriors. The Moral and Legal Status of Soldiers*, Oxford: Oxford University Press, 2010.
Schnocks, Johannes, *Das Alte Testament und die Gewalt. Studien zu göttlicher und menschlicher Gewalt in alttestamentlichen Texten und ihren Rezeptionen*, Neukirchen-Vluyn: Neukirchener Verlagsgesellschaft, 2014.
Schockenhoff, Eberhard, *Wie gewiss ist das Gewissen? Eine ethische Orientierung*, Freiburg/Breisgau: Herder, 2003.
Schockenhoff, Eberhard, *Kein Ende der Gewalt? Friedensethik für eine globalisierte Welt*, Freiburg/Breisgau: Herder, 2018.
Schotte, Dietrich, *Die Entmachtung Gottes durch den Leviathan. Thomas Hobbes über Religion*, Stuttgart: frommann-holzboog, 2013.
Schwienhorst-Schönberger, Ludger, "Recht und Gewalt im Alten Testament," in: Nadja Rossmanith et al. (eds), *Sprachen heiliger Schriften und ihre Auslegung*, 7–33, Institut für Religion und Frieden (Ethica Themen), Wien: BMLVS Heeresdruckerei, 2015.

Senghaas, Dieter, *Den Frieden denken. Si vis pacem, para pacem*, Frankfurt/Main: Suhrkamp, 1975.
Stengel, Friedemann/ Ulrich, Jörg (eds), *Kirche und Krieg. Ambivalenzen in der Theologie*, Leipzig: EVA, 2015.
Strub, Jean-Daniel/Grotefeld Stefan (eds), *Der gerechte Friede zwischen Pazifismus und gerechtem Krieg. Paradigmen der Friedensethik im Diskurs*, Stuttgart: Kohlhammer, 2007.
Stümke, Volker, *Das Friedensverständnis Martin Luthers. Grundlagen und Anwendungsbereiche seiner politischen Ethik*, Stuttgart: Kohlhammer, 2007.
Stümke, Volker/ Gillner, Matthias (eds), *Friedensethik im 20. Jahrhundert*, Stuttgart: Kohlhammer, 2011.
Svensson, Isak, *Ending Holy Wars. Religion and Conflict Resolution in Civil Wars*, Queensland: University of Queensland Press, 2012.
Tönnies, Sibylle, *Die Menschenrechtsidee. Ein abendländisches Exportgut*, Wiesbaden: Verlag für Sozialwissenschaften, 2011.
Verlage, Christopher, *Responsibility to Protect. Ein neuer Ansatz im Völkerrecht zur Verhinderung von Völkermord, Kriegsverbrechen und Verbrechen gegen die Menschlichkeit*, Tübingen: Mohr-Siebeck, 2009.
Weingardt, Markus, *Religion Macht Frieden. Das Friedenspotential von Religionen in politischen Gewaltkonflikten*, Stuttgart: Kohlhammer, 2007.
Weingardt, Markus, *Was Frieden schafft. Religiöse Friedensarbeit. Akteure, Beispiele, Methoden*, Gütersloh: Gütersloher Verlagshaus, 2014.
Weissenberg, Timo J., *Die Friedenslehre des Augustinus. Theologische Grundlagen und ethische Entfaltung*, Stuttgart: Kohlhammer, 2005.
Werkner, Ines-Jacqueline/Ebeling, Klaus (eds), *Handbuch Friedensethik*, Wiesbaden: Springer, 2017.
Werkner, Ines-Jacqueline/Meireis, Torsten (eds), *Rechtserhaltende Gewalt – eine ethische Verortung. Fragen zur Gewalt Band 2*, Wiesbaden: Springer, 2018.
Werkner, Ines-Jacqueline, *Gerechter Friede. Das fortwährende Dilemma militärischer Gewalt*, Bielefeld: transcript, 2018.

# Suggestions for Further Reading

Blustein, Jeffrey M., *Forgiveness and Remembrance. Remembering Wrongdoing in Personal and Public Life*, Oxford: Oxford University Press, 2014.
Huber, Wolfgang/ Reuter, Hans-Richard, *Friedensethik*, Stuttgart: Kohlhammer, 1990.
Margalit, Avishai, *On Compromise and Rotten Compromises*, Princeton: Princeton University Press, 2010.
Merkl, Alexander, *„Si vis pacem, para virtutes." Ein tugendethischer Beitrag zu einem Ethos der Friedfertigkeit*, Münster: Aschendorff, 2015.
von Schubert, Hartwig, *Pflugscharen und Schwerter. Plädoyer für eine realistische Friedensethik*, Leipzig: EVA, 2018.
Volf, Miroslav, *Flourishing. Why we need religion in a globalized world*, Yale: University Press, 2015.
Wink, Walter, *The Powers That Be. Theology for a New Millenium*, New York: Doubleday, 1999.

Asma Afsaruddin

# The Concept of Peace in Islam

Regardless of predominant discourses in the Western public sphere about Islam and Muslims today, it must be emphasized in this chapter that the concept of peace is a central one in Islamic thought. The Arabic word *salām* is most frequently used to connote "peace"; it shares its etymology with the name for the religion – "Islam" – itself. *As-salām* is one of the ninety-nine names for God (*Allāh*) in Islam invoked by pious Muslims everywhere; this further underscores the importance of the concept of peace in Islamic religious thought and praxis. It is also well-known that Muslims traditionally greet one another by saying "Peace be on you" (*As-sal ām ʿalaykum*) to which the response is "And peace be on you" (*Wa-ʿalaykum as-sal ām*). With regard to the general establishment of peace as a socio-organizational principle, the prevalent attitude among Muslims is that the revealed laws of God, properly interpreted and implemented, will inevitably lead to the ultimate desideratum: a just and peaceful social order.

While peace, peaceableness, and peacemaking are central concepts within Islam, the religion in its fundamental orientation cannot be described as a pacifist. Pacifism in its absolute sense is generally understood to mean an unconditional eschewal of violence under any and every circumstance.[1] In general, the Islamic moral landscape has not been receptive to the idea of pursuing non-violence as an ideological end in itself, severed from the requirement of fulfilling the conditions of social and political justice. Non-violence, after all, can be (and has been) forcibly imposed by the strong on the weak to the detriment of the latter's rights and dignity. Thus pacifism, when defined as non-violence under all circumstances and the unconditional rejection of war, even in the face of violent aggression, would be regarded in specific situations as facilitating injustice and contributing to social instability – and, therefore, morally and ethically unacceptable.

The neologism "pacificism," on the other hand, more closely encapsulates traditional Islamic attitudes towards peacemaking. Pacificism refers to a preference for peaceful conditions over war, but accepts that armed combat for defensive purposes may on occasion be necessary to advance the cause of peace.[2]

1 Jenny Teichman, *The Philosophy of War and Peace*, Exeter, UK: Imprint Academic, 2006, 171.
2 For the distinction between "pacifism" and "pacificism," see Martin Ceadel, *Pacifism in Britain 1914–1945. The Defining of a Faith*, Oxford: Clarendon Press, 1980, chap. 1; see also idem, *Thinking about Peace and War*, Oxford: Oxford University Press, 1987, 101ff.

https://doi.org/10.1515/9783110682021-003

Conditional pacifism may be another way of referring to this position. In contrast, absolute pacifism harbors the possibility of acquiescing in injustice and evil, a moral infraction that is indefensible within the Islamic ethos. The Islamic principle of *ḥisba* (enjoining good and forbidding evil) instructs that refusal to resist wrong, even if only verbally, is a grave abdication of individual and collective moral responsibility. Peace does not devolve on its own; the establishment of a non-violent social and world order requires conscious effort and constant vigilance, in addition to peaceful intent. Paradoxically, the maintenance of peace requires that those who would seek to subvert it must be resisted through a variety of peaceful means at first and ultimately through principled violence when peaceful means are exhausted. The Qur'ān uses the term *jihād* to connote this constant human struggle to promote what is essentially right and good and prevent what is evil and wrong in all aspects of life.

Discussions of peace and violence in the Islamic milieu must therefore start with a focus on the term *jihād* with its multiple, contested meanings in the various sources of Islamic thought and praxis. As always, our discussion must begin with the Qur'ān, the central revealed scripture in Islam which lays the foundation for all moral, ethical, and legal thinking in the Islamic milieu.

## 1 Jihād in the Qur'ān

*Jihād* in the Qur'ān is a broad, multivalent term and its basic signification is "struggle," "striving," "exertion." The lexeme *jihād* is frequently conjoined to the phrase "*fī sabīl Allāh*" (lit. "in the path of God") in extra-Qur'ānic literature. The full locution in Arabic, *al-jihād fī sabīl Allāh*, consequently means "struggling/striving for the sake of God." This translation points to the polysemy of the term *jihād* and its potentially different connotations in different contexts, for human striving "in the path of/for the sake of God" can be accomplished in multiple ways. A different Qur'ānic term *qitāl* specifically refers to "fighting" or "armed combat" and is a component of *jihād* in specific situations. *Ḥarb* is the Arabic word for war in general. The Qur'ān employs this last term four times: to refer to illegitimate wars fought by those who wish to spread corruption on earth (5:64); to the thick of battle between believers and non-believers (8:57; 47:4); and, in one instance, to the possibility of war waged by God and his prophet against those who would continue to practice usury (2:279).[3] This term

3 These are the only instances when the specific word *ḥarb* is employed in the Qur'ān.

is never used with the phrase "in the path of God" and has no bearing on the concept of *jihād*.

According to the Qur'ānic world-view, human beings should be constantly engaged in the basic moral endeavor of enjoining what is right and forbidding what is wrong (Qur'ān 3:104, 110, 114; 7:157; 9:71, 112, etc.). The "struggle" implicit in the application of this precept is *jihād*, properly and plainly speaking, and the endeavor is both individual and collective. The means for carrying out this struggle vary according to circumstances, and the Qur'ān often refers to those who "strive with their wealth and their selves" (*jāhadū bi-amwālihim wa-anfusihim*; for e.g., Qur'ān 8:72).

We now proceed to a discussion of the various meanings of *jihād* as they occur in the Qur'ān. Although not intended as an exhaustive discussion, our survey below brings to the fore the different inflections of *jihād* against the backdrop of some of the key events in the life of the Prophet Muhammad (d. 11/632).

## 1.1 The Meccan Period (610 CE–1/622)

According to Islamic sources, the Prophet Muhammad began to receive revelations roughly around 610 CE. This constitutes the beginning of the Meccan phase of Muhammad's prophetic career which lasted until the famous *hijra* or emigration to Medina in 622 CE, which corresponds to the first year of the Islamic calendar. During the Meccan period, the Muslims were not given divine permission to physically retaliate against the pagan Meccans who persecuted them for their profession of monotheism and instituted several harsh measures against them – including an economic boycott, forced starvation, and physical torture. Verses revealed in this period counsel the Muslims to steadfastly endure the hostility of the Meccans while continuing to practice and propagate their religion. Although the Qur'ān recognizes the right to self-defense for those who are wronged, it maintains in this early period that to bear patiently the wrong-doing of others and to forgive those who cause them harm is the superior course of action in resisting evil. A cluster of verses (42:40–43) reveal this highly significant, non-militant dimension of struggling against wrong-doing (and, therefore, of *jihād*) in this early phase of Muhammad's prophetic career. These verses state:

> The requital of evil is an evil similar to it: hence, whoever pardons and makes peace, his reward rests with God – for indeed, He does not love evil-doers. Yet surely, as for those who defend themselves after having been wronged – no blame whatever attaches to them: blame attaches but to those who oppress people and behave outrageously on

> earth, offending against all right; for them is grievous suffering in store! But if one is patient in adversity and forgives, this is indeed the best resolution of affairs.[4]

In the Qur'ānic discourse, patience is thus a component and a manifestation of the *jihād* of the righteous; quietist and activist resistance to wrong-doing are equally valorized. One Qur'ānic verse thus (16:110) states "As for those who after persecution fled their homes and strove actively (*jāhadū*) and were patient (*ṣabarū*) to the last, your Lord will be forgiving and merciful to them on the day when every soul will come pleading for itself." Another (47:31) states, "We shall put you to the test until We know the active strivers (*al-mujāhidīn*) and the quietly forbearing (*aṣ-ṣābirīn*) among you." Quietist, non-violent struggle is not the same as passivity, however, which when displayed in the face of grave oppression and injustice, is clearly marked as immoral in the Qur'ānic view. "Those who are passive" (Ar. *al-qā'idūn* in Arabic) earn divine rebuke in the Qur'ān (4:95).

Furthermore, generous posthumous rewards are promised for the conscious inculcation of patience. For instance, Qur'ān 39:10 states that "those who are patient will be given their reward without measure;" and Qur'ān 25:75 states "They will be awarded the high place [in heaven] for what they bore in patience . . . abiding there forever."

Probably the most important verse in the Qur'ān that extols the attribute of patient forbearance is Qur'ān 3:200. This verse states, "O those who believe, be patient and forbearing (*iṣbirū*), outdo others in forbearance (*ṣābirū*), be firm, and revere God so that you may succeed." If we look at a cross-section of early and late exegeses on this critical verse, we find a range of scholarly opinions on the meanings of this verse and their ramifications.

In his brief commentary on this early Medinan verse,[5] the early Meccan exegete Mujāhid b. Jabr al-Makkī (d. 104/722)[6] attributes to his contemporary al-Ḥasan al-Baṣrī [d. 110/728] the comment that the verse counsels believers to be

4 Q 42:40–43. All translations are mine, although I have freely consulted existing translations.

5 Most commentators are agreed that Chapter 47 (Sūrat Muḥammad) is early Medinan, revealed during the first and second years of the *hijra*, with the exception of verse 18 which is regarded as having been revealed during the *hijra*. A few exegetes considered the entire chapter to be late Meccan.

6 Mujāhid was born in Mecca and was a student of Ibn 'Abbās. He transmitted from 'Alī b. Abī Tālib, Ubayy b. Ka'b, and 'Abd Allāh b. 'Umar; cf. Fuat Sezgin, *Geschichte des arabischen Schrifttums*, Leiden: Brill, 1967; (henceforth abbreviated as GAS), 1:29; and the article "Mudjāhid b. Djabr al-Makkī", in *Encyclopaedia of Islam*, new edition, ed. H. A. R. Gibb, Leiden: Brill, 1960–2003[2], 7:293.

steadfast in their religion and bear patiently with the unbelievers until they despair of their religion and to be firm against the polytheists.[7]

About a generation later, Muqātil b. Sulaymān (d. 150/767) comments that the verse exhorts believers to be steadfast in carrying out the commands and duties imposed by God; to be forbearing along with the Prophet wherever he is; and to be firm against the enemy in the path of God (*wa-rābiṭū l-ʿaduw fī sabīl allāh*) "until they forsake their religion for yours" and to fear God and not be disobedient. Those who do that will be successful (*fa-qad aflaḥa*).[8] Despite the addition of *fī sabīl allāh*, more commonly used with derivatives of *jihād* in the Medinan period, there is no hint of military combat in engaging the enemy in this manner. Through the deliberate conjoining of *fī sabīl allāh* to *ribāṭ* in this context, Muqātil establishes a link between patient forbearance in difficult circumstances and non-violent striving in the path of God against his enemies.

The celebrated exegete Muḥammad b. Jarīr aṭ-Ṭabarī (d. 310/923) offers us a much more detailed commentary on Qurʾān 3:200 and records reports attributed to different early authorities that nicely encapsulate for us the contested meanings of this verse by the late third/ninth century. A number of earlier exegetes, as he documents, understood the verse to mean, "Remain steadfast in your faith and bear patiently the [harm caused by] unbelievers and be firm with them." Among this group of exegetes were the Successors al-Ḥasan al-Baṣrī, Qatāda b. Diʿāma (d. 118/736) and Ibn Jurayj (d. 150/767).[9]

Another group of exegetes maintained that the verse means, "Remain steadfast in your faith, and wait patiently for My promise to you regarding your obedience to Me, and be firm against your enemies." Muḥammad b. Kaʿb al-Quraẓī (d. 118/736)[10] thus explained this verse as, "Remain steadfast in your religion, await patiently [the fulfillment] of my promise to you, and be firm against My and your enemies until they forsake their religion for yours."[11] In these two exegetical clusters, it is noteworthy that the believers are being exhorted to practice non-violent patient forbearance in the face of the persecution inflicted on them by the pagan Meccans.

---

**7** Mujāhid, *Tafsīr Mujāhid*, ed. ʿAbd ar-Raḥmān aṭ-Ṭāhir b. Muḥammad as-Surtā, Islamabad, n.d., 44.

**8** Muqatil b. Sulayman, *Tafsīr*, ed. ʿAbd Allāh Maḥmūd Shiḥāta, Beirut: Muʾassasat at-taʾrīkh alʿarabī 2002, 1:324. Muqātil then goes on to append details of the letter that Muhammad wrote to the Christians of Najrān and ʿUmar's affirmation of it, details of which need not detain us here.

**9** Aṭ-Tabarī, *Jāmiʿ al-bayān ʿan taʾwīl āy al-Qurʾān*, Beirut, 1995, 3:561–62.

**10** Muḥammad b. Kaʿb b. Sulaym al-Quraẓī, the oldest among the Tābiʿūn, composed a Qurʾān commentary; cf. *GAS*, 1:32.

**11** Aṭ-Ṭabarī, *Tafsīr*, 3:562.

As for the Arabic imperative *rābiṭū*, in the verse, many of the early commentators understand it to mean "Observe your prayers," that is to say, "wait for them [the prayers] one after another." One report attributed to Shuraḥbīl [b. Saʿīd] (d. ca. 123/740),[12] as recorded by aṭ-Ṭabarī in his commentary, relates that ʿAlī [b. Abī Ṭālib] had remarked that the Prophet on one occasion asked, "Shall I not indicate to you what will cause God to blot out your sins and lapses? The traces of ablution upon the limbs and anticipation of the prayers, one after another – that is *ar-ribāṭ*."[13] Yet another report from Shuraḥbīl relates from the Companion Jābir b. ʿAbd Allāh (d. 78/697) that the Prophet had asked, "Should I not indicate to you the means whereby God effaces mistakes and pardons sins?" When his gathered Companions answered in the affirmative, he replied, "The reaching of the ablution [water] to its [proper] places; frequenting of the mosques, and waiting for prayer one after another; that is *ribāṭ* for you." Two slightly variant reports attributed to Abū Hurayra are also recorded by aṭ-Ṭabarī.[14]

These early non-violent understandings of *rābiṭū* stand in sharp contrast to the exegeses of later commentators who frequently connect this imperative with the post-prophetic military activity of *ribāṭ* that refers to the guarding of frontier territories on horseback. This contestation is indicated in a report that emanates from Abū Salama b. ʿAbd ar-Raḥmān (d. between 94–104/712–722)[15] who once asked his nephew, "Do you know what this verse was revealed about?" When the boy answered in the negative, Abū Salama said, "My nephew, there were no campaigns (*ghazw*) during the time of the Prophet which required guarding the frontiers (*yurābaṭ fīhi*). It rather means to be vigilant concerning the [performance of the] prayers, one after the other."

In his explication of Qur'ān 3:200, ar-Rāzī painstakingly describes the various inflections of patience, necessitated by the fact that human existence is

12 Shuraḥbīl was one of the earliest writers of *maghāzī*, and, according to Sufyān b. ʿUyayna, the most knowledgeable of it. Ibn Isḥāq and al-Wāqidī did not transmit from him, but Ibn Saʿd related from him concerning the *hijra* of the Prophet; cf. GAS, 1:279.

13 Aṭ-Ṭabarī, *Jāmiʿ*, 3:562.

14 Ibid, 3:562–63.

15 Abū Salama b. ʿAbd ar-Raḥmān b. ʿAwf b. ʿAbd ʿAwf az-Zuhrī l-Madanī was known by his *kunya* (patronymic); his given name may have been either ʿAbd Allāh or Ismāʿīl. He is said to have narrated *ḥadīth*s from prominent Companions such as ʿUthmān b. ʿAffān, Ṭalḥa (although some doubted this), Abū Hurayra, ʿĀ'isha and Umm Salama, while his son ʿUmar, ʿUrwa b. az-Zubayr, az-Zuhrī, and Mūsā b. ʿUqba transmitted from him. Ibn Saʿd mentions him in his *Ṭabaqāt* as one of the second generation Medinans, although he is said to have met Muhammad. Ibn Saʿd describes him as a "trustworthy, learned man, who transmitted many *ḥadīth*s" (*thiqa faqīh kathīr al-ḥadīth*); Abū Zurʿa concurred; see Ibn Ḥajar, *Tahdhīb*, 6:351–53.

comprised of two spheres: one which has to do with the individual only, and the other which is shared with others. In the first sphere, humans must practice patience and steadfastness (*aṣ-ṣabr*) [in relation to themselves and their duties], while in the second they must practice forbearance (*al-muṣābara*) vis-à-vis others. Patience (*ṣabr*) is of various kinds, continues ar-Rāzī. Firstly, one must have patience in learning the intricacies of viewpoints and complexities of proofs when acquiring knowledge of God's unity, justice, prophecy, and resurrection, and in deriving categorical conclusions from conflicting and dubious arguments. Secondly, one must have patience in undertaking the difficulties inherent in carrying out religious obligations and recommended actions. Thirdly, one must be patient and steadfast in refraining from all prohibited things. Fourthly, patience is to be exercised in the face of hardships in this world and its afflictions in the form of disease, poverty, hunger, and fear. All these categories are subsumed under his command *iṣbirū*, while each of these three categories include other situations without restriction.[16]

As for *muṣābara*, it refers to forbearance during unpleasant occurrences between oneself and others. This includes showing forbearance towards members of one's family, neighbors, and relatives when they behave badly. *Muṣābara* also includes refraining from taking revenge on those who cause you harm, as counseled in Qur'ān 7:199, "Turn away from the ignorant." Also relevant in this context is the verse, "When they pass by frivolity, they do so with dignity" (Qur'ān 25:72). Further included in *muṣābara* is showing a preference and love for the other over oneself, as mentioned in Qur'ān 59:9, "They prefer [others] over themselves even when reaping poverty." Thus, *muṣābara* also consists of forgiving those who wrong/oppress you (*al-ʿafu ʿamman ẓalamaka*); ar-Rāzi references Qur'ān 2:237 here which states, "Forgiveness is closer to God-conscious piety" (*lit-taqwā*). Moreover, *muṣābara* includes commanding the good and preventing wrong. Since the one undertaking this duty may face harm, it includes fighting in self-defense. It includes as well showing forbearance with the foolish in trying to assuage their doubts and dispel their inclination towards falsehood. It is thus categorically established that *iṣbirū* has to do with the individual while *ṣābirū* deals with interactions between the individual and others.[17]

Ar-Rāzī warns, however, that even when practicing patience and forbearance, humans are still subject to the baser aspects of their nature, such as anger and covetousness which impel them to act in contrary ways. Thus, humans are engaged throughout their lives in striving to combat and overcome these base

16 Ar-Rāzī, *At-Tafsīr al-kabīr*, Beirut: Dār iḥyā' at-turāth al-ʿarabī, 1999, 3:473.
17 Ibid.

desires, which cause them to lose patience and sacrifice forbearance – hence they are counseled to hold firm (*wa-rābiṭū*). Since this striving is an act like other acts which occur in response to an incentive and objective, the incentive and objective of this striving is fear of and reverence for God (*at-taqwā*) in order to attain success and salvation (*al-falāḥ wa-n-najāh*). To summarize this rather lengthy section, ar-Rāzī says *al-murābaṭa* refers to the effort expended in controlling and suppressing all manner of evil propensities (*al-quwa*) within the human self and *al-falāḥ* (success) is the end result for those who thereby prefer to please God out of reverence for him over submission to their base desires.[18]

Like ar-Rāzī, the seventh/thirteenth century Andalusian commentator al-Qurṭubī (d. 671/1273) affirms that Qur'ān 3:200 encompasses within it much counsel concerning a range of issues – from how to present oneself in this world to the enemy to attaining the felicity of the next world. It urges believers to be steadfast in their obligations and in fighting their base desires, for *ṣabr* is self-restraint. As for *muṣābara*, some said its meaning is to be firm with the enemy, as maintained by Zayd b. Aslam, whereas al-Ḥasan al-Baṣrī said it means steadfastness in observing the five prayers. Others said it means to persist in acting contrary to the incitements of one's lower self, "for it beckons and he resists." According to ʿAṭā' [b. Abī Rabāḥ, d. 733/115] and the previously mentioned al-Quraẓī, it means to wait patiently for the fulfillment of the promise made to humans [by God] and to not despair subsequently, waiting instead for deliverance and relief (*al-faraj*) [after their hardship]. Abū ʿUmar[19] would say, "Waiting for deliverance with patience is an act of worship."[20]

Al-Qurṭubī, refers approvingly to Abū Salama's commentary – well-known to us by now – that the verb *rābiṭū* refers to waiting for the daily prayers, especially since the practice of manning the frontiers did not exist during the Prophet's time. Like aṭ-Ṭabarī and other exegetes, al-Qurṭubī notes that Abū Salama had also related the *ḥadīth* in which Muhammad repeats three times that *ribāṭ* means waiting for the prayers one after another, while acts associated with prayer, such as ablution and going to the mosque, wipe out one's sins.[21]

To underscore this meaning of *ribāṭ*, al-Qurṭubī refers to other *ḥadīth*s attesting that *ribāṭ* has also been understood to mean the patient anticipation and observance of the daily prayers. For example, Abū Nuʿaym al-Ḥāfiẓ[22] recorded a

---

**18** Ibid., 3:473–74.

**19** It is not clear which Abū ʿUmar this is.

**20** Al-Qurtụbī, *Al-Jāmiʿ li-aḥkām al-qur'ān*, Beirut: Dār al-kitāb al-ʿarabī, 2001, 4:313–14.

**21** Ibid., 4:314.

**22** This is Abū Nuʿaym Aḥmad b. ʿAbd Allāh b. Mihrān al-Iṣbahānī, the author of *Ḥilyat al-awliyā' wa-ṭabaqāt al-ashfiyā'*. He is generally regarded as a reliable transmitter of *ḥadīth*; see

*ḥadīth* from ʿAbd Allāh b. ʿUmar who said that the Prophet prayed the Maghrib prayer one evening with some of his Companions, after which some withdrew and some stayed on in the mosque. When Muhammad returned to the mosque right before theʿIshāʾ prayer later in the evening, he found nineteen people already there. This made him point to the heavens with his index finger and exclaim, "Rejoice, O gathering of Muslims, your Lord has opened up a door in the heavens and boasted about you to the angels, saying, 'O my angels, look at these servants of mine who have just performed a religious duty and are now waiting for the next one'!" The report makes clear that religious merit, if God should so will, may also accrue to the individual who practices *ribāṭ* in this manner, comments al-Qurṭubī.[23]

It is clear from these exegetical discussions of Qurʾān 3:200 that *ṣabr*/*muṣābara* along with *ribāṭ*/*murābaṭa* all refer to various aspects of the internal, spiritual struggle which is a constant aspect of striving in the path of God, that is to say of *al-jihād fī sabīl allāh*. In later literature this internal, spiritual struggle is renamed *jihād an-nafs* or *al-jihād al-akbar*. The later provenance of this terminology (probably after the fourth/tenth century) has led a number of Orientalists and modern polemicists to mistakenly assert that no Qurʾānic genealogy can be found for the concept of the internal spiritual and moral struggle. Furthermore, according to this group, its assumed extra-Qurʾānic development establishes that *jihād* is fundamentally a militant concept in the Qurʾān.[24] Nothing could be further from the truth of course. As is clearly evident from our discussion, the Qurʾān commands humans to constantly struggle to vanquish the base temptations of the lower self and to thereby achieve greater moral heights and establish peaceful relations with one's fellow human beings. This exhortation is conveyed through commands to practice patient forbearance, to pray constantly, and to forgive the failings of others. The inability of these Orientalists and polemicists to understand this Qurʾānic imperative to practice *ṣabr* as an essential component of *jihād* broadly defined has allowed them to create the dangerous and self-serving myth of Islam as fundamentally a militant religion.

---

Ibn Khallikān, *Wafayāt al-aʿyān wa-anbāʾ abnāʾ az-zamān*, ed. Iḥsān ʿAbbās, Beirut, n.d., 1:37; Ibn Qāḍī Shuhba, *Ṭabaqāt ash-shāfiʿiyya*, Hyderabad, 1398/1978, 1:201–2; adh-Dhahabī, *Mīzān al-iʿtidāl fī naqd ar-rijāl*, ed. Badr ad-Dīn an-Naʿsānī, Cairo, 1325/1907, 1:52.

**23** Al-Qurṭubī, *Jāmiʿ*, 4:318.

**24** Representatives of this group include Alfred Morabia, *Le Ğihâd dans l'Islam medieval. Le "combat sacré" des origines au XIIe siècle*, Paris: Albin Michel, 1993; Emile Tyan, art. "Djihād," in *Encyclopaedia of Islam*, ed. C.E. Bosworth et al., new ed., supplement Leiden: Brill, 1980–1992), 2: 538–539; and, more recently, Andrew Bostom, *The Legacy of Jihad. Islamic Holy War and the Fate of Non-Muslims*, New York: Prometheus Books, 2005, and David Cook, *Understanding Jihad*, Berkeley: University of California Press, 2005.

## 1.2 Jihād as Justified Fighting in the Qur'ān

To continue our delineation of the semantic topography of *jihād* in the Qur'ān, we must now turn our attention to the combative component (*qitāl*) nestled within this umbrella term. The question we must ask is when, where, why, and how does *qitāl* become a required component of striving in the path of God, according to the Qur'ān? To answer this question, we have to focus on the Medinan phase of the Prophet Muhammad's career and take a close look at select critical verses which deal with the necessity of fighting under certain conditions as a moral duty imposed upon believers.

### 1.2.1 Jihād in the Medinan period

During the ten years of the Meccan period and the first two years of the Medinan period (622–624 CE) – for a total of roughly twelve years – Muslims were not allowed to physically retaliate against their pagan Meccan persecutors. But shortly after the emigration to Medina, a specific Qur'ānic verse (22:39–40) that permitted fighting for the first time was revealed. The verse states:

> Permission [to fight] is given to those against whom fighting has been initiated, and indeed, God has the power to help them: those who have been driven from their homes against all right for no other reason than their saying, "Our Provider is God!" For, if God had not enabled people to defend themselves against one another, monasteries, churches, synagogues, and mosques – in all of which God's name is abundantly glorified – would surely have been destroyed.

In the Meccan period as we recall, Qur'ān 42:40–43 (mentioned above) allowed self-defense but not through violent means; the reasons for undertaking this kind of non-violent self-defense are the wrongful conduct of the enemy and their oppressive and immoral behavior on earth. In Qur'ān 22:39–40, two more reasons are given: the initiation of fighting by the enemy and wrongful expulsion of people from their homes for no other reason than their affirmation of belief in one God. Furthermore, the Qur'ān asserts, if people were not allowed to defend themselves against aggressive wrong-doers, all the houses of worship – it is noteworthy that Jewish and Christian places of worship are included alongside Muslim ones – would be destroyed and thus the word of God extinguished. It is reasonable to infer from this verse that Muslims may resort to defensive combat even on behalf of non-Muslim believers who are the object of the hostility of non-believers. These are the just causes for which Muslims may go to war against an intractable enemy, like the pagan Meccans, against whom all peaceful means of resistance were deployed and exhausted during the preceding non-violent twelve years period before divine permission was finally granted to fight in self-defense.

Another verse (Qur'ān 2:217) states:

> They ask you concerning fighting in the prohibited months.[25] Answer them: "To fight therein is a serious offence. But to restrain [people] from following the cause of God, to deny God, to violate the sanctity of the sacred mosque, and to expel its people from its environs is with God a greater wrong than fighting in the forbidden month. [For] disorder and oppression are worse than killing.

In this verse, the Qur'ān acknowledges the enormity of fighting during the prohibited months but at the same time stresses the higher moral imperative of maintaining order and resisting wrong-doing. Therefore, when both just cause and righteous intention exist, war in self-defense against an intractable enemy may become obligatory.

The Qur'ān further states that it is the duty of Muslims to defend those who are oppressed and who call out to them for help (4:75), except against a people with whom the Muslims have concluded a treaty (8:72). The Qur'ān also counsels (5:8), "Let not rancor towards others cause you to incline to wrong and depart from justice. Be just; that is closer to piety." This verse therefore warns against succumbing to unprincipled and vengeful desire to punish and inflict disproportionate damage.

The principle of proportionality is in fact emphasized in Qur'ān 2:194 where it is explicitly stated: "Whoever attacks you attack him to the extent of his attack. Fear God and know that God is with the God-fearing." Aṭ-Ṭabarī helpfully points to a range of interpretive opinions among the exegetes. According to Ibn 'Abbās, this verse was revealed in Mecca when the Muslims were few in number and too weak to subdue the polytheists, who would revile and physically hurt them. Thus, God allowed the Muslims to retaliate to the extent to which they were hurt or to be patient or to forgive, the latter being the ideal response, according to Ibn 'Abbās. When the Prophet emigrated to Medina, and God increased him in strength, and rescued Muslims from their victimhood and gave them control of their own affairs, he commanded them not to attack one another like the people of the pre-Islamic period.[26]

But others maintained, continues aṭ-Ṭabarī, that the verse was Medinan and allowed believers to fight those among the polytheists who fought them.

---

**25** These were four specific months deemed sacred in the pre-Islamic period during which fighting was prohibited. These months are: Shawwāl, Dhū 'l-Qa'da, Dhū 'l-Ḥijja, and Muḥarram.
**26** Aṭ-Ṭabarī, *Jāmi'*, 2:205.

Those who subscribed to this interpretation included Mujāhid b. Jabr with which interpretation aṭ-Ṭabarī agrees, since the verses preceding Qur'ān 2:194 have to do with fighting the unbelievers, which was allowed only after the *hijra*. The remainder of the verse is also Medinan since fighting was not permitted in the Meccan period. The meaning of "Those who attack you retaliate against them to the extent to which they attack you," may be compared to "Fight in the path of God those who fight you." The resulting meaning is that "whoever attacks you in the Kaʿba and fights you, attack and fight him to the extent of his act of aggression," according to the law of *talionis* (*qiṣāṣ*). Others have maintained that the meaning of this verse is, "Whoever aggresses against you – that is, whoever is hostile towards you and inflicts a wrong – you may attack him – that is inflict harm on him to the same extent – in exact retribution (*qiṣāṣ*) for what he did to you, without transgressing the limits (*lā ẓulman*)."[27] The verse concludes, aṭ-Ṭabarī continues, with an assurance to the believers that those who adhere to these limits are the pious ones and God is with the pious who revere him, carry out the religious obligations, and avoid what is forbidden.[28]

### 1.2.2 Initiation of Hostilities

The Qur'ān also has specific injunctions with regard to initiation of hostilities. Qur'ān 2:190 which reads, "Fight in the cause of God those who fight you, but do not commit aggression, for God loves not aggressors," forbids Muslims from commencing hostilities. Fighting must be in response to a prior act of aggression by the enemy.

Qur'ān 2:190 is one of the most significant verses concerning the combative *jihād* in the Medinan period. Early exegetes, such as Mujāhid and as-Suddī (d. 128/745), unequivocally subscribed to the view that the verse explicitly forbids Muslims from ever initiating aggression against anyone, including obvious wrongdoers/oppressors (*az-ẓālimīn*), in any place, sacred or profane. Thus Mujāhid comments that according to this verse, one should not fight until the other side commences fighting.[29] According to Muqātil, this verse is specifically a denunciation of the Meccans who had commenced hostilities at al-Hudaybiyya (6/628), leading to a repeal of the prohibition imposed upon Muslims against fighting near

27 Ibid., 2:205–06.
28 Ibid., 2:206.
29 Mujāhid, *Tafsīr*, 23.

the Ka'ba. "Do not commit aggression" and "God does not love aggressors" is an indictment, he asserts, against the Meccans who began to fight during the sacred month in the sacred sanctuary, which was a clear act of aggression (*fa-innahu 'udwān*). Permission to engage the pagan Meccans in fighting was clearly contingent, according to Muqātil, upon their having initiated hostilities, which abrogates the earlier prohibition against fighting in the sanctuary. "If they desist" means that if they should cease fighting and acknowledge the oneness of God, then God will forgive their previous adherence to polytheism and show mercy towards them "in Islam."[30]

So unambiguous is the Qur'ānic proscription against initiating fighting in these verses – as was stressed by a majority of scholars prior to the late third/ninth century – that several later exegetes had to resort to abrogation as a hermeneutic tool to nullify this explicit command. We see this process underway in aṭ-Ṭabarī's commentary. Even though aṭ-Ṭabarī records the views of the early exegetes Mujāhid and as-Suddī and acknowledges that Qur'ān 2:190 (and the next four verses) centered on the events at al-Ḥudaybiyya sets specific, strict limitations on armed combat for the faithful, he nevertheless proceeds to nullify a number of these restrictions by considering them abrogated by verses occurring in different chapters referring to escalated hostilities in different contexts (Badr, Uḥud, etc.). Earlier authorities cited by aṭ-Ṭabarī himself, such as the Successors ar-Rabī' and 'Ikrima, did not however declare any of the verses in this cluster to have been abrogated, particularly by later verses from the ninth chapter, as would become fairly common in exegetical works from after the fourth/tenth century. This process, however, was not ineluctable. The famous Mu'tazilī exegete az-Zamakhsharī (d. 538/1144) continues to maintain in the first half of the sixth/twelfth century that Muslims may not ever initiate fighting on the basis of Qur'ān 2:190. However, he also maintains – by referencing Qur'ān 9:36 at this point – that once the polytheists have initiated aggression, Muslims may as a consequence fight against all of them. Ar-Rāzī in the early sixth/twelfth century similarly upholds the general non-aggression clause so that Muslims may not ever initiate fighting, but once fighting has commenced against the pagans, then it must be continued until they abandon both their polytheism and aggression, for one presupposed the other. Al-Qurṭubī expresses similar views. The progressive understanding of *fitna* and *ẓulm* as specifically referring to "polytheism" and/or "unbelief" rather than broadly to "trials" and "wrongdoing" facilitated this exegetical strand.[31]

**30** Ibid, 1:167–68.

**31** For an extensive discussion of the exegeses of Qur'ān 2:190–194, see Asma Afsaruddin, *Striving in the Path of God. Jihad and Martyrdom in Islamic Thought*, Oxford: Oxford University Press, 2013, 43–58.

### 1.2.3 Qur'ān 9:12–13 and 2:216

Qur'ān 9:12–13 is another important cluster of verses providing a list of reasons that make physical retaliation permissible against an enemy. These verses state:

> If they break their pacts (*aymānahum*) after having concluded them and revile your religion, then fight the leaders of unbelief. Will you not fight a people (*qawman*) who violated their oaths and had intended to expel the Messenger and commenced [hostilities] against you the first time?

There is a general unanimity among the exegetes that these verses underscore the prior aggression of the pagan Meccans against the Muslims which necessitated fighting against them. The early exegetes Mujāhid, Muqātil and the Ibāḍī exegete of the late third/ninth century Ibn Muḥakkam (d. 290/903) understand these verses as allowing Muslims to fight those polytheists who had violated their pacts (*aymānahum*) with them, had denigrated Islam, and initiated hostilities against them.[32] Muqātil specifically remarks that Qur'ān 9:12 refers to the Meccan polytheists who violated their agreement with the Prophet to desist from fighting from two years. Instead of observing the truce, they secretly armed the clan of Kināna and goaded the latter to attack the clan of Khuzāʿa, who had made peace with Muhammad. As a result, "the Prophet, peace and blessings be upon him, deemed it licit to fight them" (*fa-staḥalla an-nabī ṣallā allāhu ʿalayhi wa-sallam qitālahum*). The next verse refers to the same incident concerning Kināna and Khuzāʿa, he continues, and denounces the Quraysh for descending on *Dār an-nadwa*,[33] conspiring to either kill the Prophet, shackle him, or expel him. The Quraysh commenced hostilities when they marched to Badr to fight the Muslims.[34]

In his commentary on Qur'ān 9:12, aṭ-Ṭabarī says that it is a critique of those among the Quraysh who violated the terms of their pact with Muhammad that they would not fight the Muslims nor provide aid to their enemies; additionally, these Qurayshīs had defamed Islam. The leaders of the unbelievers thus had to be fought against so as to cause them to desist from providing aid to the enemies of Muslims and from reviling Islam. Aṭ-Ṭabarī notes that most exegetes agree with this interpretation, although they differ on the precise

---

32 Ibid. 58–59.

33 *Dār al-nadwa*, referred to a kind of town hall in Mecca to the north of the Kaʿba where the Quraysh used to meet to deliberate upon important communal matters and hold certain public events; see the ar. "Dār al-nadwa" in the *Encyclopaedia of Islam*, second edition, ed. Peri Bearman et al., published online 2012.

34 Muqātil, *Tafsīr*, 2:159–60.

identification of the leaders of the unbelievers.[35] The next verse adds to the polytheists' list of misdeeds their expulsion of the Prophet and their commencement of hostilities during the battle of Badr. Others, like Mujāhid, say it is because they began to fight the clan of Khuzāʿa who were the allies of Muhammad. The verse concludes, comments aṭ-Ṭabarī, by warning that Muslims should fear God more than they do the polytheists.[36]

In the late sixth/twelfth century, ar-Rāzī notably emphasizes that the verb *bada'ūkum* in Qur'ān 9:13 draws attention to the fact that the aggressor is unequivocally the greater offender (*tanbīhan ʿalā anna l-bādiʾ aẓlam*).[37] Along with Qur'ān 2:190, Qur'ān 9:13 is understood generally to offer the most explicit iteration of this scriptural condition – "that they had initiated aggression against you" (*wa-hum bada'ūkum awwala marratin*) – for resorting to armed combat.

Those who would prefer to infer an unending divine command to fight non-Muslims qua non-Muslims look elsewhere in the Qur'ān. One of their favorite verses is Qur'ān 2:216 which states "Fighting has been prescribed for you even though you find it displeasing. Perhaps you dislike something in which there is good for you and perhaps you find pleasing that which causes you harm. But God knows and you do not."

A diachronic survey of the exegeses of Qur'ān 2:216 however unearths an early critical position on fighting that became nearly completely forgotten or ignored in the later period. The verse which describes fighting as a prescribed duty (*kutiba ʿalaykum al-qitāl*) instigated a discussion among exegetes as to who exactly was being addressed in the pronominal suffix – *kum*. Aṭ-Ṭabarī provides valuable documentation that early Medinan authorities like Ibn Jurayj and ʿAṭā' b. Abī Rabāḥ believed that only the Muslims during the time of Muhammad are the referent in this verse. Ar-Rāzī adds to this list the name of another Medinan scholar ʿAbd Allāh b. ʿUmar who had similarly maintained that the duty of fighting was imposed on the Companions alone. Contrasted to these Medinan scholars is the early Syrian authority Makḥūl who is said to have sworn at the Kaʿba that fighting (he uses the word *ghazw*, no doubt to set up a contrast to defensive fighting) was obligatory. His student, the well-known Syrian jurist al-Awzāʿī (d. 15/774), was more equivocal – and pragmatic – in his views, as reported by aṭ-Ṭabarī. In the fifth/eleventh century, al-Wāḥidī is on record as endorsing the early position that fighting as a religiously prescribed duty applied only to the Companions, citing ʿAṭā' b. Abī Rabāḥ as his authority.

**35** For a discussion of the various possibilities, see aṭ-Ṭābarī, *Jāmiʿ*, 6:329.

**36** Ibid., 6:331.

**37** Ibid., 5:535.

It is clear, therefore, on the basis of this substantial documentation that this position was hardly a minority and negligible one in Islamic history. As late as the fifth/eleventh century, our sources indicate that this remained a credible and dominant view subscribed to by influential scholars and that there was considerable resistance on the part of some scholars to the attempts of other scholars to aggrandize the status of fighting as a religious duty incumbent on *all* believers for all time.[38]

There is a clear regional breakdown by the time of the Successors in the Umayyad period – our survey indicates that Ḥijāzī scholars, prominently among them ʿAṭāʾ b. Abī Rabāḥ, tended to understand Qurʾān 2:216 as mandating fighting only for the Companions during the time of Muḥammad while Syrian and, one assumes, generally pro-Umayyad jurists, like Makḥūl, subscribed to the view that the verse contained a general commandment for all eligible Muslims to fight. This latter position is articulated most explicitly by ar-Rāzī in the early sixth/twelfth century when he asserts that "in spite of what ʿAṭāʾ said," the verse in its use of *ʿalaykum* is to be understood as imposing the duty of fighting on both those who were present at the time of its revelation and those who will come later. Given his defensive tone, ar-Rāzī is fully aware that he is going against the prevailing exegetical near-consensus of his time that, according to this verse, fighting as an individual religious obligation had lapsed after the time of the Prophet. But clearly the legal sensibilities of his time and the historical exigencies during the Seljuq period plagued by vulnerability to external enemies must have prompted ar-Rāzī to adopt this line of reasoning. Al-Qurṭubī in the last quarter of the seventh/thirteenth century hews to very similar views; in his case, his concern to establish fighting as an individual duty on the basis of this verse is prompted by the precarious situation in which Muslims in al-Andalus find themselves in the face of the Reconquista. It should be noted in this context that other early authorities like Ibn ʿAṭiyya and Sufyān ath-Thawrī cited by our exegetes construed the military *jihād* in general as a voluntary (*taṭawwuʿ*) and collective act.[39]

### 1.2.4 Qurʾān 9:5 and Qurʾān 9:29

Roughly by the century with the maturation of the legal schools (*madhāhib*), we start to detect a strong exegetical tendency to infer a general mandate from

38 See a discussion of these various views in Afsaruddin, *Striving in the Path of God*, 65–71.
39 Ibid.

the Qur'ān to fight offensive or expansionist wars. Two Medinan verses are often cited by many jurists as setting up a religious obligation to fight non-Muslims until they convert to Islam or at least capitulate to Muslim rule. The first is Qur'ān 9:5 which states: "When the sacred months have lapsed, then slay the polytheists (*al-mushrikīn*) wherever you may encounter them. Seize them and encircle them and lie in wait for them. But if they repent and perform the prayer and give the *zakāt*, then let them go on their way, for God is forgiving and merciful." The second is Qur'ān 9:29, which states: "Fight those who do not believe in God nor in the Last Day and do not forbid what God and his messenger have forbidden and do not follow the religion of truth from among those who were given the Book until they proffer the *jizya* with [their] hands in humility."

A survey of exegetical works reveals that until the Seljuq period, the first verse Qur'ān 9:5 was not the subject of much attention among Qur'ān commentators. Early exegetes understood the *mushrikīn* mentioned in this verse to refer specifically to those polytheists with whom the first generation of Muslims did not have pacts and therefore historically circumscribed in its application. Thus, Muqātil understands the *mushrikīn* mentioned in the verse as a reference to those polytheists during the time of Muhammad with whom there was no pact (*ʿahd*) and who represented a hostile faction against whom the Muslims could legitimately fight.[40]

Aṭ-Ṭabarī also understands the verse to command the slaying of hostile polytheists specifically during the time of Muhammad.[41] After aṭ-Ṭabarī it is noteworthy that al-Wāḥidī and az-Zamakhsharī both pay scant attention to Qur'ān 9:5, since the verse was unambiguously understood by them to refer to the treatment of the polytheists during Muhammad's time and thus to have no further applicability in their own time and place.[42]

It is extremely significant that no exegete in our survey refers specifically to Qur'ān 9:5 as the *āyat as-sayf* ("verse of the sword") before the late eighth/fourteenth century. We first encounter this designation in our survey in the commentary of the well-known exegete Ibn Kathīr (d. 774/1373) from the Mamluk era.[43] Ibn Kathīr's characterization of this verse indicates to us that by the Mamluk period when Islamic realms were under continuous assault by the Crusaders and the Mongols, many scholars felt impelled to derive a general expansive mandate from Qur'ān 9:5 and other such historically circumscribed

**40** Muqātil, *Tafsīr*, 2:157.

**41** Ibid.

**42** See this discussion in Afsaruddin, *Striving in the Path of God*, 72–73.

**43** Ibn Kathīr, *Tafsīr al-Qur'ān al-ʿaẓīm*, Beirut: Dār al-Jil, 1990, 2:322.

Qur'ānic verses to fight and punish all those who posed a threat to the well-being of Muslims.

Another verse Qur'ān 9:29 has been understood by a number of scholars to mandate endless warfare against the People of the Book unless they convert to Islam and that it too, like Qur'ān 9:5, abrogates other conciliatory verses in the Qur'ān. A diachronic survey of the exegeses of this verse reveals the following: Qur'ān 9:29 was understood by the early exegete Mujāhid b. Jabr as a specific reference to the battle of Tabūk,[44] thereby implying that the scriptuaries referenced in this verse are specifically hostile factions from among them, like the Byzantine Christians.[45] However, later exegetes understand this verse as referring broadly to Jews and Christians who are required to humbly pay the *jizya* as a marker of their inferior legal status vis-à-vis Muslims. Aṭ-Ṭabarī also acknowledges that the historical context for the revelation of this verse was war with Byzantium, and soon thereafter Muhammad undertook the campaign of Tabūk, as maintained by Mujāhid and others.[46]

Ar-Rāzī in the late sixth/twelfth century helpfully preserves a spectrum of opinions among Muslims scholars on how to interpret key locutions in the verse which indicate divine dissatisfaction with certain contingents from among the People of the Book. By his time, most exegetes read this verse as containing a blanket condemnation of Jews and Christians because they do not believe as Muslims do. Ar-Rāzī however documents the important views of an early Kufan exegete Abū Rawq (ʿAṭiyya b. al-Ḥārith al-Hamadānī al-Kūfī, d. 140/757) who stated that this verse chides Jews and Christians for not heeding the prescriptions contained in the Torah and the Gospel respectively. Abū Rawq's views are more credibly in line with several Qur'ānic verses (Qur'ān 5:44–47; 5:66) which call upon Jews and Christians to follow the Torah and the Gospel respectively, and other verses which refer to different revealed laws and ways of life existing concurrently with Islam without having been abrogated (for example, Qur'ān 5:48), and which affirm that previous revelations are confirmed, rather than superseded, by the Qur'ān (for example, 2:89,101; 5:48; 10:37). No doubt over time as Muslims became majorities outside of the Arabian peninsula and began to develop a growing sense of communal solidarity vis-à-vis non-Muslims, often against the backdrop of continuing skirmishes with the Byzantine Christians

44 Mujāhid, *Tafsīr*, 99.
45 This verse refers specifically to the Byzantines who are said to have amassed their forces on the Syrian border in preparation for an attack on Muslims in the year 630. Arabic sources refer to the event as the Battle of Tabūk, although no battle was eventually fought since the Byzantine forces failed to materialize.
46 Aṭ-Ṭabarī's, *Jāmiʿ*, 6:349.

through the Umayyad and ʿAbbasid periods and eventual bloody encounters with the Crusaders from the fifth/eleventh century onward, most exegetes preferred to construe the disobedience of the People of the Book referred to in Qurʾān 9:29 as their disobedience to the laws of Islam rather than their own laws – *against* the overall thrust of the Qurʾān itself.

One notable exception to this general trend was al-Qurṭubī; on the issue of treatment of the *ahl adh-dhimma*, he notably includes the following reports as a caveat. According to Muslim b. Ḥajjāj, the Companion Hishām b. Ḥakīm b. Ḥizām[47] was once walking by a group of Nabateans in Syria who had been made to stand in the sun and the order to "pour oil on their heads" was given. When Hishām inquired into their case, he was told that they had withheld the *jizya*. Hishām remonstrated that he had heard the Prophet say, "Indeed God will punish those who punish humans in the world." Hishām then went to see ʿUmayr b. Saʿd, the ruler at that time in Palestine, and narrated this *ḥadīth* before him, and ʿUmayr gave the order for them to be released. Al-Qurṭubī continues by saying that "our scholars" [sc. jurists in Andalusia] were of the view that if the *jizya* was withheld by the *ahl adh-dhimma* despite being able to pay it, then it was a punishable offense, but if they were incapable of making the payment, then it was not permissible to punish them. Inability to pay the *jizya* nullified the requirement to do so, he continues; the rich may not remit it on behalf of the poor. Al-Qurṭubī concludes this section by relating another *ḥadīth*, as transmitted by Abū Dāʾūd from Ṣafwān b. Salīm (alive 132/749) "from a number of the children of the Companions of the Messenger of God, peace and blessings be upon him, from their fathers," which relates that Muhammad had said, "Whoever oppresses the one who has entered into a pact [with Muslims; *muʿāhidan*] or disparages him or imposes on him a responsibility beyond his ability or takes something from him with animus (*bi-ghayr ṭayyib an-nafs*), then I will be his (sc. the *dhimmī*'s) advocate on the Day of Judgment."[48]

Al-Qurṭubī's comments are clearly intended to provide an important corrective to the more discriminatory and harsh attitudes that had surfaced among a number of Muslim authorities by his time, as reflected in the exegetical and juridical literature of the time. In spite of the perilous times in which he lived, al-Qurt .ubī's own multi-faith environment in Muslim Spain seems to have predisposed him to invoke more irenic and humane *ḥadīth*s commanding kindness towards

47 For whom see, for example, Ibn Ḥajar, *Tahdhīb at-tahdhīb*, ed. Khalīl Maʾmūn Shīḥā et al., Beirut: Dār al-maʿrifa, 1996, 6:26–27.
48 Al-Qurtụbī, Jāmiʿ, 8:106.

the *dhimmī* in the Muslim's protection,[49] in contradistinction to practically all the exegetes surveyed above who were anxious to articulate the superior legal and confessional status of Muslims in their more religiously segregated societies. He articulated these views even in the face of the Reconquista. Evidently, in the multifaith environment of Andalusia which had historically allowed for the coexistence of Jews, Christians, and Muslims, al-Qurṭubī had learned to distinguish between peaceful and hostile members of these monotheistic communities.

## 2 Qurʾānic Ethics of Refraining from Fighting and Peacemaking

On the topic of the military *jihād*, scholars in general have paid far scantier attention to verses that call for the cessation of fighting and which thereby create a moral imperative to make peace. A holistic reading of the Qurʾān establishes that in addition to being defensive, fighting in the Qurʾān is clearly limited in nature. The Qurʾānic ethics of desisting from fighting and making peace are just as important as the rules it sets down for conducting a justified war. Thus Qurʾān 60:7–9 state:

> Perhaps God will place affection between you and those who are your enemies for God is powerful and God is forgiving and merciful. God does not forbid you from being kind and equitable to those who have neither made war on you on account of your religion nor driven you from your homes; indeed God loves those who are equitable. God forbids you, however, from making common cause with those who fight you on account of your religion and evict you from your homes and who support [others] in driving you out. Those who make common cause with them are wrong-doers.

Another important verse that commands peaceful relations with non-Muslims who display no hostility towards Muslims is Qurʾān 9:6 which states: "If anyone from among the polytheists asks you for protection, grant it to him so that he may hear the word/speech of God (*kalām Allāh*), then escort him to a place of safety for him. That is so because they are a people without knowledge." This verse immediately following the so-called "sword verse" appears almost startlingly dissonant because it countermands the seemingly absolute, general injunction to "slay the polytheists wherever you may find them" in the preceding

49 Al-Qurṭubī is clearly making a distinction between peaceful Christians willing to live with Muslims in al-Andalus and the hostile Christian invaders descending from the north during his time.

verse. In seeming contrast, this verse advocates courteous, even deferential, behavior towards the same Arab polytheists who however show a willingness to learn about the Qurʾān and evince no hostility towards Islam and Muslims per se.

These verses, therefore, make very clear that Muslims may fight only those who have clearly aggressed against them and persecuted them for their faith. Non-Muslims who live peacefully with them and display no hostility are to be treated kindly and equitably, regardless of what they choose to believe. These perspectives are reflected in the *tafsīr* literature starting from the early to the late pre-modern period. A majority of the exegetes that I surveyed in an earlier study affirm that both sets of verses – Qurʾān 60:7–9 and 9:6 – remain unabrogated and their injunctions to treat peaceful non-Muslims justly and kindly remain valid for all times.[50] This position is powerfully expressed by aṭ-Ṭabarī who affirms that the most appropriate exegesis of Qurʾān 60:7–9 is as follows: God has not forbidden Muslims from acting kindly and fairly with all those from any and every religion and creed who do not fight them and do not expel them from their homes. Aṭ-Ṭabarī also summarily dismisses the suggestion that this verse is abrogated, for it clearly permits the faithful to be kind to "the people of war" (*ahl al-ḥarb*), whether blood-relatives or not, who bear no ill-will towards Muslims and as long as such relationships do not compromise the security of Muslims. For God loves those who are equitable (*al-munṣifīn*), who give people their due rights, are personally just to them, and do good to those who are good to them. As for Qurʾān 60:9, it forbids believers from helping and befriending those from among the unbelievers in Mecca who fight them over religion and evict them from their homes. Those who do so are wrong-doers and violate the command of God; this was the view of Mujāhid.[51]

However, a number of early exegetes like Muqātil b. Sulaymān, ʿAbd ar-Razzāq (d. 211/827), and Ibn Muḥakkam (d. 290/903) were of the view that Qurʾān 60:7–9 had been abrogated by Qurʾān 9:5. Exegetes surveyed from aṭ-Ṭabarī onwards however are forceful in continuing to affirm the unabrogated status of Qurʾān 60:7–9. These later exegetes invoke early authorities like Ibn ʿAbbās, Mujāhid, Muqātil b. Ḥayyān, and al-Kalbī in support of their position.[52]

Another early and highly important exegetical divide becomes apparent in our discussion concerning Qurʾān 9:6: aḍ-Ḍaḥḥāk and as-Suddī are said to

**50** Afsaruddin, *Striving in the Path of God*, 82–90.
**51** Aṭ-Ṭabarī, *Jāmiʿ*, 12:63.
**52** Afsaruddin, *Striving in the Path of God*, 82–87.

have maintained its abrogation by Qur'ān 9:5, while Mujāhid and al-Ḥasan al-Baṣrī affirmed its continuing validity. The early Shī'ī exegete Furāt unusually maintains that Qur'ān 9:6 had abrogated Qur'ān 9:5. The overwhelming majority of exegetes – from Mujāhid to al-Qurṭubī – uphold the commandment in Qur'ān 9:6 to offer safe conduct to peaceful non-Muslims *as an enduring and binding one.*[53]

The quintessential Qur'ānic verse concerning peacemaking is Qur'ān 8:61, which states, "And if they should incline to peace, then incline to it [yourself] and place your trust in God; for he is all-hearing and all-knowing." Aṭ-Ṭabarī says that God in this verse addresses the Prophet and counsels him that if he should fear treachery and perfidy on the part of a group of [unspecified] people (*qawm*), then he should withdraw from them and fight them. But "if they should incline to making peace with you and abandon warfare (*wa-in mālū ilā musālamatika wa-mutārakatika al-ḥarb*), either through entry into Islam, or payment of the *jizya*, or through the establishment of friendly relations (*muwāda'a*), then you should do the same for the sake of peace and peacemaking (*min aṣbāb as-silm wa-ṣ-ṣulḥ*)."[54] Those who have supported this exegesis include Qatāda who glossed *as-silm* as *aṣ-ṣulḥ* but who also maintained that this verse had been abrogated by Qur'ān 9:5 and 9:36.[55] The Successor Ibn Zayd (d. 182/798) was of the opinion that the verse commanded making peace with the opposite side (*fa-ṣāliḥhum*) but then the military *jihād* abrogated this commandment.[56]

At this point, aṭ-Ṭabarī weighs in and comments that Qatāda's statement, echoed by others to the effect that this verse had been abrogated, cannot be supported on the basis of the Qur'ān, the *sunna*, or reason. An abrogating verse, he continues, is one that nullifies the injunction/prescription contained in an abrogated verse in every aspect. If it does not meet this essential criterion, then it cannot function as an abrogating verse. Qur'ān 9:5 cannot abrogate Qur'ān 8:61 because the latter concerns the Qurayẓa clan who were Jews and therefore one of the *ahl al-kitāb*. God has permitted believers, he says, to make peace with the People of the Book and abandon fighting them when *jizya* is taken from them. Qur'ān 9:5 on the other hand has to do *only* with Arab polytheist idolaters from whom *jizya* cannot be taken. Neither verse invalidates the injunction contained in the other and both remain unabrogated (*muḥkam*) concerning its specific content. He notes that Mujāhid had identified the referent

**53** Ibid., 88–90.
**54** Aṭ-Ṭabarī, *Jāmi'*, 6:278.
**55** Ibid.
**56** Ibid., 6:278.

as the Banū Qurayẓa in this verse.[57] The rest of the verse, concludes aṭ-Ṭabarī, assures Muhammad that he need only place his faith in God when making peace with "the enemies of God," for he hears and knows all that transpires during such negotiations.[58] Aṭ-Ṭabarī's unequivocal defense of the unabrogated status of Qur'ān 8:61 and the reasons he advances for rejecting Qur'ān 9:5 as an abrogating verse in this case are particularly noteworthy.

Aṭ-Ṭabarī views on the continuing applicability of this verse beyond the generation of the Prophet were repeated by a number of exegetes after him: for example, ar-Rāzī in the late sixth/twelfth century says that after the preceding verse (Qur'ān 8:60) exhorted Muslims to assemble their forces against the enemy, should the same enemy incline to peace, then Muslims are commanded to accept their offer of peace (*fa-l-ḥukm qubūl aṣ-ṣulḥ*) in Qur'ān 8:61. Furthermore, this injunction to make peace with the enemy is a broad and general one, not restricted to the generation of the first Muslims.[59]

All these verses taken together–Qur'ān 9:6, 60:7–9, and 8:61–clearly establish that the military *jihād* in the Qur'ān is most categorically **not** holy war that is fought to **impose** a religion, since it may be undertaken only in **defense** of a religious community that is already under attack by hostile forces, mainly for professing their (monotheistic) faith. As a collectivity, these verses make clear that fighting may continue only as long as the adversary engages in fighting and that Muslims must resort to peaceful arbitration when the other side sues for one. The invocation of *naskh* or abrogation by some scholars in regard to these verses testifies to the ingenuity of a belligerent faction which was determined to find scriptural sanction for war that may be fought on ostensible theological grounds and thus offensively, thereby overriding explicit Qur'ānic restrictions on resorting to fighting. Such hermeneutic legerdemain conveniently provided a justification for the wars of conquest that ensued after the Prophet's death, whose worldly impetus was only too apparent to certain pietist groups.

## 3 Negotiating the Polyvalence of the Term *Jihād*

The literature from the first three centuries of Islam reveals that there were competing definitions of how best to strive in the path of God, engendered by the polyvalence of the term *jihād* as occurs in the Qur'ān. Recent rigorous research

---

**57** Ibid., 6:278–79.
**58** Ibid., 6:279.
**59** Ar-Rāzī, *Tafsīr*, 5:500–01.

has established that there was a clear divergence of opinions regarding the nature of *jihād* and its imposition as a religious duty on the believer through the first century of Islam and into the second half of the second century. In an excellent article on early, competing conceptions of *jihād*, two distinguished historians of Islamic thought, Roy Mottahedeh and Ridwan as-Sayyid, have pointed out that during the Umayyad period, there were multiple and conflicting perspectives on this subject held by jurists from Syria, for example, who were close to ruling circles and jurists from the Hijaz, who were outside the orbit of such circles.[60] Early jurists not aligned with official circles, like Sufyān ath-Thawri (d. 161/778), were of the opinion that *jihād* was primarily defensive, and that only the defensive *jihād* may be considered obligatory on the individual. Jurists from the Hijaz (from the province of western Arabia that includes Mecca and Medina) generally tended to place greater emphasis on religious practices such as prayer and mosque attendance and did not consider *jihād* obligatory for all. On the other hand, Syrian jurists like al-Awzāʿī (d. 157/773), who were close to the Umayyads, held the view that even aggressive war may be considered obligatory. No doubt this last group was influenced by the fact that the Syrian Umayyads during this time were engaged in border warfare with the Byzantines and there was a perceived need to justify these hostilities on a theological and legal basis.[61] It would not be an exaggeration to state that expressing support for expansionist war at this time (the Umayyad period) was to proclaim one's support for the existing government and its policies.

By the early Abbasid period, roughly mid-late second/eighth century, the military aspect of *jihād* appears to have become foregrounded over other spiritual and non-militant significations of this term in juridical and official circles. *Jihād* from this period on would progressively be conflated with *qitāl* ("fighting"), collapsing the distinction that the Qurʾān maintains between the two. As the jurists and religious scholars of all stripes became consolidated as a scholarly class and accrued to themselves commensurate religious authority by the fourth/tenth century, they arrogated to themselves the right to authoritatively define *jihād* and circumscribe the range of activities prescribed by it. With the powerful theory of abrogation *(naskh)* at their disposal, some of the jurists effectively rendered null and void the positive injunctions contained in the Qurʾānic verses which explicitly permitted the conclusion of truces with foes

---

**60** Mottahedeh, Roy/al-Sayyid, Ridwan, "The Idea of the *Jihad* in Islam before the Crusades," in: Angeliki E. Laio/ Roy Parviz Mottahedeh (eds), *The Crusades from the Perspective of Byzantium and the Muslim World*, 23–29, Washington, D. C., 2001.
**61** Ibid, 25–27.

and counseled peaceful co-existence with particularly the "People of the Book." One of the most important verses declared by a number (by no means all) of these scholars to have been abrogated or superseded is Qur'ān 2:256, which forbids compulsion in religion, by verses which give the command to fight.[62]

This, however, does not mean that these scholars henceforth regarded coercion of non-believers to embrace Islam as valid.[63] But since Qur'ān 2:256 had been adduced as a proof-text by those who inveighed against the concept of an offensive *jihād* – for that might lead the way to coercive conversions – the opposite camp felt impelled to declare the injunction contained in this verse (and other verses which advocated peaceful, non-militant relations with non-Muslims) to be abrogated or at least superseded by other verses, such as Qur'ān 9:5[64] or Qur'ān 9:73.[65] In the opinion of these jurists, such an act of abrogation or supersession would remove scripture-based objections to the waging of offensive battles in order to extend the political realm of Islam. Once conquered, non-Muslims could be given the choice of either embracing Islam or paying the *jizya*. Usually compliance with the second option (in the absence of any desire to convert) meant that the third option posited by this camp of jurists for non-believers in the event of non-compliance – "to be put to the sword" – was unlikely to be exercised; and the jurists in all probability envisioned that this is how matters would turn out. Thus, even these jurists could not have conceived of *jihād* primarily as "holy war" to effect the conversion of non-Muslims. They could in fact be accused of doing just the opposite – that is, of politicizing and **secularizing** *jihād* so that it could be deployed as "expansionist war," which could be launched to further the state's imperial objectives to expand territorially and extend its political dominion. The motives for effecting this transformation were clearly more imperial than empyreal!

The monovalent and aggressive understanding of *jihād* promoted by such scholars within the context of international relations undermined the rich diversity of meanings associated with the term in Qur'ānic and early *ḥadīth* discourse. The jurist Muḥammad b. Idrīs ash-Shāfi'ī (d. 204/820) is said to have

62 For an extensive discussion of this controversial topic, see aṭ-Ṭabarī, *Jāmi'*, 3:15–19; also 'Abd ar-Raḥmān ibn al-Jawzī, *Nawāsikh al-Qur'ān*, Beirut: Dār al-kutub al-'ilmiyya, n.d., 93–94.

63 Aṭ-Ṭabarī, *Jāmi'*, 3:18.

64 Ibn al-Jawzī, *Nawāsikh*, 93.

65 This verse states: "Strive against the unbelievers and the hypocrites and be stern with them; their refuge is Gehenna, a wretched destiny;" see Ibn al-'Arabī, *an-Nāsikh wa-l-mansūkh fī l-qur'ān al-karīm*, Beirut: Dār al-kutub al-'ilmiyya, 1997, 61; Ibn al-Jawzi, *Nawāsikh*, 94.

been the first to permit *jihād* to be launched against non-Muslims as offensive warfare, although he qualified non-Muslims as referring only to pagan Arabs and not to non-Arab non-Muslims. He further divided the world into *dār al-isl ām* ("the abode of Islam") and *dār al-ḥarb* ("the abode of war," referring to non-Muslim territories), while allowing for a third possibility *dār al-ʿahd* ("the abode of treaty") or *dār aṣ-ṣulḥ* ("the abode of reconciliation"), to which abode non-Islamic states which had entered into a peace treaty with the Islamic state by rendering an annual tribute may be admitted.[66] In the absence of actual hostilities, the Shāfiʿī school of thought posited an existing state of "cold war" between the abodes of Islam and war, which required constant vigilance on the part of the former against the latter.[67] Political theorists after ash-Shāfiʿī would enshrine this concept in their writings by averring that one of the duties of the caliph was to launch *jihād* at least once a year; although others were of the opinion that this duty could be fulfilled by simply being in an adequate state of military preparedness to forestall enemy attacks.[68]

Ash-Shāfiʿī 's perspective on *jihād* was, in many ways, a marked departure from earlier juristic thinking and reflects a certain hardening of attitudes towards non-Islamic polities by his time (late second/eighth and early third/ninth century). This is quite evident when his views are compared with those of jurists from the earlier Ḥanafī school of law, eponymously founded by Abū Ḥanīfa (d. 150/767). Ḥanafi jurists, for example, did not subscribe to a third abode of treaty, as devised by ash-Shāfiʿī, but were of the opinion that the inhabitants of a territory which had concluded a truce with the Muslims and paid tribute to the latter became part of the abode of Islam and entitled to the protection of the Islamic government.[69] The Ḥanafīs also adhered to the position that non-believers could only be fought if they resorted to armed conflict, and not simply on account of their disbelief.[70] This remained a principle of contention between later Shāfiʿī and Ḥanafī jurists.

It is worth emphasizing that the concepts of *dār al-islām* and *dār al-ḥarb* have no basis in the Qurʾān nor in the *sunna*. The invention of these terms and

---

**66** Ash-Shāfiʿī, *Kitāb al-umm*; Bulaq: al-Maktaba al-kubrā l-amīriyya, 1903, 4:103–4.

**67** Ash-Shāfiʿī, *ar-Risāla*, ed. Aḥmad Shākir, n. pl., 1891, 430–32

**68** Majid Khadduri, *War and Peace in the Law of Islam*, Baltimore: Johns Hopkins University Press, 1955, 64–65.

**69** See *The Islamic Law of Nations. Shaybani's Siyar*, tr. and ed. Majid Khadduri, Baltimore: Johns Hopkins University Press, 1966, 12–13; Khadduri, *War and Peace*, 145.

**70** As did the Ḥanafī jurist Aḥmad aṭ-Ṭahāwī (d. 933) in his *Kitāb al-Mukhtaṣar*, ed. Abū l-Waf ā l-Afghānī, Hyderabad: Lajnat iḥyāʾ al-maʿārif al-nuʿmāniyya, 1950, 281; cited by Khadduri, *Islamic Law*, 58.

the resulting aggrandizement of the military aspect of *jihād* were based rather on Realpolitik in the Abbasid period. An imperial state, such as the Abbasid one, in control of a vast and diverse political realm, had to develop a sophisticated law of nations, termed in Arabic *as-siyar* (lit. "motions" "travels"). From the vantage point of Realpolitik, *jihād* could be understood as defensive or offensive fighting whose primary purpose was to guarantee the legitimate security needs of the polity. Thus, during the Umayyad period (661–750 CE), the constant border skirmishes with the hostile Byzantines predisposed Syrian jurists in particular to endorse the concept of an offensive *jihād* (in contrast to jurists living far away from the metropole) as an effective military strategy against an intractable enemy.

In the process of politicizing *jihād*, the daring abrogation of particularly the critical Qurʾānic verse 2:256, which states "There is no compulsion in religion," as maintained by some scholars, was by no means accepted by all. For example, aṭ-Ṭabarī resolutely maintained that this verse had not been abrogated and its injunction remained valid for all time.[71] Furthermore, aṭ-Ṭabarī in his juridical work *Ikhtilāf al-fuqahāʾ* does not list Qurʾān 9:5 as an abrogating verse, revealing that even as late as the fourth/tenth century, there was by no means a juridical consensus on the status of this verse.[72]

## 4 Patience (*Ṣabr*) as an Aspect of *Jihād* in Ḥadīth and Edifying Literature

To resurrect the early competing perspectives on what constitutes *jihād* we have to turn to certain sources which have preserved for us a broader semantic and historical trajectory of these two concepts. The belligerent statist construal of *jihād* which had become ascendant by the third/ninth century did not efface the earlier multiplicity of views regarding the term's meaning. Some sources lead one to the belief that by this century, the "hawkish" factions had basically won and the non-militant "dovish" faction had receded to the sidelines at best, or worse, completely superseded. But scrutiny of alternative sources at our disposal, alternative that is to standard juridical, exegetical, and *ḥadīth* literature, allows one to maintain that dissenting, pietistic circles continued to register

71 Aṭ-Ṭabarī, *Jāmiʿ*, 3:25.

72 Aṭ-Ṭabarī, *Ikhtilāf al-fuqahāʾ*, Cairo: n.p., 1933, 1–21.

their opposition to the state's unseemly glorification of armed combat in service of the polity, couched in strident religious rhetoric. Such "dissidents" were not necessarily pacifists or mystic-ascetics (this period is well before the rise of institutionalized Sufism) but pious ethicists who were scandalized by the state's perceived willful disregard for the Qur'ān's concern to establish just cause for military combat.

The "literature of dissent" arising in these circles constitutes in part a genre called *faḍā'il aṣ-ṣabr,* which means "the excellences or virtues of patience." It is a genre that is meant to be contraposed to the well-known genre of *faḍā'il al-jihād*, which praises the excellences or merits of armed combat.[73] The *faḍā'il aṣ-ṣabr* literature extols the virtue of patience, which is an important aspect of *jihād* as struggle against wrong-doing; this is clearly evident from the Qur'ān as we have established so far. Taking its cue from the Qur'ān, the two genres – excellences of fighting vs. patience – represent countervailing and competing definitions of how best to struggle for the sake of God. A third/ninth century work, for instance, on the merits of patience by Ibn Abī d-Dunyā (d. 281/894), called *aṣ-Ṣabr wa-th-thawāb ʿalayhi* records the following report on the authority of ʿIṣma Abī Ḥukayma, who related,

> The Messenger of God, peace and blessings be upon him, wept and we asked him, "What has caused you to weep, O Messenger of God?" He replied, "I reflected on the last of my community and the tribulations they will face. But the patient from among them who arrives will be given the reward of two martyrs (*shahīdayn*)."[74]

This report categorically challenges other, perhaps better known reports which assign the greatest merit to military martyrs and posits instead a different, non-militant understanding of virtuous self-sacrifice. Such a self-sacrificing patient individual will be given the reward of two military martyrs, affirming the higher moral status of this individual. Scrutiny of such *faḍā'il aṣ-ṣabr* reports yield valuable information as to what kinds of groups and individuals were circulating such reports and their objectives for doing so.[75]

---

73 The *faḍā'il* literature is a prolific one in the medieval period. Works of praise were composed about many meritorious activities and religious duties, in addition to prominent people and places. For a brief introduction to this genre, see Asma Afsaruddin, *Excellence and Precedence. Medieval Islamic Discourse on Legitimate Leadership*, Leiden: Brill, 2002, 26–35, and the references cited therein.

74 See Ibn Abī d-Dunyā, *As-Ṣabr wa-th-thawāb ʿalayhi*, Beirut: Dār Ibn Ḥazm, 1997, 85.

75 See Afsaruddin, *Striving in the Path of God*, 190–204.

## 5 Militant vs. Non-Militant Struggle: A Contest of Piety

Many of the *faḍā'il aṣ-ṣabr* reports[76] contained in Ibn Abī d-Dunyā's work referred to above testify in fact to a competitive discourse on piety that emphasizes the primacy of the Qur'ānic virtues of patience and forbearance over other traits and activities, including the military *jihād*. One such laudatory report is attributed to a certain Abū ʿImrān al-Jūnī who stated, "After faith, the believer (*ʿabd*) has not been given anything more meritorious (*afḍal*) than patience with the exception of gratitude, but it (sc. patience) is more meritorious of the two and the fastest of the two to reap recompense (*thawāb*) [for the believer]."[77] A similar report attributed to the second/eighth century scholar Sufyān b. ʿUyayna (d. 196/811) says, "The believers (*al-ʿibād*) have not been given anything better or more meritorious than patience, by means of which they enter heaven."[78]

These reports are at odds with other, more frequently quoted reports which claim that falling on the battlefield brings swift and immeasurable heavenly rewards to the martyr. One of the best known reports on the issue of compensation for the *shahīd* is recorded by Muslim b. al-Ḥajjāj (d. 261/875) and Ibn Māja (d. 273/887) in their two authoritative *ḥadīth* collections which states that all the sins of the military martyr will be forgiven except for his debt.[79]

A report in the early collection of *ḥadīth* by ʿAbd ar-Razzāq aṣ-Ṣanʿānī, referred to earlier, is attributed to the famous second/eighth century preacher and scholar al-Ḥasan al-Baṣrī (d. 110/728), in which he relates that the Prophet had stated, "Embarking upon the path of God or returning from it is better than all the world and what it contains. Indeed, when one of you stands within the battle ranks, then that is better than the worship of a man for sixty years."[80] ʿAbd ar-Razzāq records another report in which a certain Abū Mujliz relates that he was passing by a Qur'ān reciter, who was reciting the verse "God has favored those who strive with their wealth and their selves by conferring on

**76** See further Asma Afsaruddin, "Competing Perspectives on Jihad and Martyrdom in Early Islamic Sources," in *Witnesses to Faith?. Martyrdom in Christianity and Islam*, ed. Brian Wicker, Aldershot, U.K.: Ashgate Publishing, 2006, 15–31.

**77** Ibn Abī d-Dunyā, *Ṣabr*, 85.

**78** Ibid., 51.

**79** See A. J. Wensinck, *Concordance et indices la tradition musulmane*, Leiden: Brill, 1988, 2:165.

**80** ʿAbd ar-Razzāq aṣ-Ṣanʿānī. *Al-Muṣannaf*, ed. Ayman Naṣr ad-Dīn al-Azharī, Beirut: Dār al-kutub al-ʿilmiyya, 2000, 5:170.

them a rank above those who are sedentary" (4:95). At this point Abū Mujliz interrupted the reciter by saying, "Stop. It has reached me that it is seventy ranks, and between each two levels, [a distance of] 70 years is reserved for the emaciated charge horse."[81] Abū Mujliz's extemporaneous commentary provides evidence that, firstly, *jihād* has come to be understood by a number of influential people in the second/eighth century to primarily indicate armed combat, undermining the term's Qur'ānic polysemy, and, secondly, that such activity had come to be highly rated as a religiously mandated duty to be rewarded by generous divine recompense in the hereafter.

When we turn to the excellences of patience literature, it is noteworthy that Ibn Abī d-Dunyā records a report in praise of patience which invokes language very similar to that of Abū Mujliz's report in praise of *jihād*. Considered together, these two reports suggest that there were efforts made to counter this kind of excessive glorification of the merits of military activity. This statement is attributed to the fourth caliph ʿAlī b. Abī Ṭālib (d. 40/661), who relates:

The Messenger of God, peace and blessings be upon him, said, "Patience is of three kinds: patience during tribulations; patience in obedience to God, and patience in avoiding sin. Whoever has patience during a tribulation until he averts it by the seemliness of his forbearance, God will ordain for him three hundred levels [of recompense]; the distance between each level would equal that between the sky and the earth. And whoever has patience in obedience to God, God writes down for him six hundred levels; the distance between each level would equal that between the boundaries of the earth till the edge of the divine throne. And whoever has patience in avoiding sin, God prescribes for him nine hundred levels; the distance between each level is twice the distance between the boundaries of the earth up to the edge of the divine throne."[82]

The similar idiom contained in these two reports in terms of how many levels or ranks the armed combatant and the patient quietist would earn or rise to in the hereafter suggests the vaunting nature of these reports and their conscious positing of opposed hierarchies of moral excellence.

Another well-known *ḥadīth* quotes the Prophet as remarking on his return from a military campaign, "We have returned from the lesser *jihād* (sc. physical, external struggle) to the greater *jihād* (sc. spiritual, internal struggle)."[83] This latter *ḥadīth* underscores the two principal modes of carrying out *jihād* and a

---

**81** Ibid.

**82** Ibn Abī d-Dunyā, *Ṣabr*, 31.

**83** This *ḥadīth*, which appears to have emanated from Sufi circles, is recorded by al-Ghazālī in "The book of invocation," *Iḥyāʾ ʿulūm ad-diīn*, translated by Kojiro Nakamura as *Ghazali on Prayer*, Tokyo: University of Tokyo, 1975, 167. For further attestations of this *ḥadīth*, see John

hierarchical ordering of their merits, with the internal, spiritual struggle trumping the external, physical one. This *ḥadīth* is not to be found in the early collections, but its advocacy of the superiority of the spiritual struggle is reflected in another prophetic statement found in the relatively early *ḥadīth* works of Ahmad ibn Hanbal (d. 241/855) and at-Tirmidhi (d. 279/892), which states, "One who strives against his own self is a *mujahid*, that is, carries out *jihād*."[84] Another *ḥadīth* recorded by Muslim similarly emphasizes the internal, spiritual aspect of striving for God; it affirms, "Whoever strives (*jāhada*) with his heart is a believer."[85] The existence of such reports testify to the concurrent existence of spiritual and non-militant dimensions of *jihād* with the militant from an early period. They additionally serve as valuable proof-texts for undermining the polemical assertions of many today that there is no early evidence for *jihād* as spiritual struggle and that such spiritual significations represent a later development.

## 6 Privileging Non-Militancy

The multiple, non-militant significations of the phrase *fī sabīl Allāh*, particularly in the early period, is clear from a noteworthy report recorded in ʿAbd ar-Razz āq's *Muṣannaf*, which relates that a number of the Companions were sitting with the Prophet when a man from the tribe of Quraysh, apparently a pagan and of muscular build, came into view. Some of those gathered exclaimed, "How strong this man looks! If only he would expend his strength in the way of God!" The Prophet asked, "Do you think only someone who is killed [sc. in battle] is engaged in the way of God?" He continued, "Whoever goes out in the world seeking licit work to support his family is on the path of God; whoever goes out in the world seeking licit work to support himself is on the path of God. Whoever goes out seeking worldly increase (*at-takāthur*) has gone down the path of the devil" (*fa-huwa fī sabīl al-shayṭān*).[86] This report contains a rebuttal to those who would understand "striving in the way of God" in primarily military terms. It extols instead the quotidian struggle of the individual to live his or her life "in the way of God" (*fī sabīl Allāh*), which infuses even the most mundane of licit activities with moral and spiritual significance and thus earning divine approbation.

---

Renard, "*Al-Jihad al-Akbar*: Notes on a Theme in Islamic Spirituality," *Muslim World* 78 (1988), 225–42.

**84** Wensinck, *Concordance*, 1:389.

**85** Ibid., 5:455.

**86** ʿAbd ar-Razzāq, *Muṣannaf*, 5:272.

The report also emphasizes the importance of personal intention in determining the moral worth of an individual's act. Since the meritorious nature of an individual's striving for the sake of God is contingent upon purity of intent, one may understand this report as counseling caution against accepting at face value ostentatious pietism or assuming that what appears to be a pious activity to humans will be deemed as such by God who alone can know the true intention of the individual. Correct *niyya* or intention determines the moral valence of an act, according to the famous *ḥadīth*, "Actions are judged by their intentions."

Other reports proclaim that those practicing the virtues of veracity and patience and evincing compassion for the disadvantaged are equivalent in moral status to the military martyrs and are striving equally hard in the path of God. A report recorded by Ibn Abī d-Dunyā attributed to ʿAbd al-ʿAzīz b. Abī Rawwād (d. 156/775), a pious *mawlā* (a non-Arab Muslim convert) of Khurasanian descent, relates, "A statement affirming the truth (*al-qawl bi-l-ḥaqq*) and patience in abiding by it is equivalent to the deeds of the martyrs."[87] Three of the most authoritative Sunnī *ḥadīth* compilers – al-Bukhārī, Muslim, and at-Tirmidhī (d. 280/893) – record a report in which the Prophet declared that "one who helps widows and the poor are like fighters in the path of God."[88]

Over time, these non-militant construals of *jihād* became even more entrenched and came to include a number of other morally laudable pursuits which contribute to the moral excellence of the individual and enhance the welfare of the people. The fifth/eleventh century Andalusian scholar Ibn Ḥazm (d. 456/1064) affirmed a general higher moral valuation of the defense of Islam through non-militant, verbal and scholarly means over militant ones by a hierarchical ordering of actions which qualify as meritorious struggle in the path of God. Thus, he says, *jihād* is best exercised, in order of importance, through a) invitation of people to God by means of the tongue; b) the defense of Islam through sound judgment and carefully-considered opinions, and c) through armed combat. With regard to the third type of *jihād*, Ibn Hazm states that this is its least important aspect. When we look at the Prophet himself, he says, we realize that the majority of his actions fall into the first two categories, and although he was the most courageous of all human beings, he engaged in little physical combat.[89] This hierarchy clearly challenges the mainly

---

**87** Ibn Abī d-Dunyā, *Ṣabr*, 116.

**88** Wensinck, *Concordance*, 1:389.

**89** Ibn Ḥazm, *Kitab al-Fiṣal fī l-milal wa-l-ahwāʾ wa-n-niḥal*, Cairo: Muʾassasat al-khānjī, 1903, 4:135.

juridical understanding of *jihād* as primarily military activity and affirms that the struggle of the learned scholar in explaining and defending Islam through reasoned argument and the marshaling of rational proofs is far more meritorious. Discourses such as these emanating from the scholarly classes which stress the acquisition and dissemination of knowledge as the best form of *jihād* quite clearly challenge and undermine exclusively statist and militaristic definitions of it propagated in administrative circles.

# 7 Peace in the Qur'ān and Ḥadīth

## 7.1 Peace in the Qur'ān

So far we have described the Qur'ān's overall perspective on issues of war and peace as twin concepts that are related to one another. Discussion of *jihād*, especially in its military aspects, may lead to the perception that the Qur'ān places undue emphasis on physical and military resistance to wrongdoing and aggression. This would be far from the truth – peace (*as-salām* and its derivatives) is idealized as the state that governs the life of the faithful and defines their relationship with the Creator and his creation. Within the Qur'ānic worldview, peace is the ultimate desideratum, the achievement of which is the most noble and loftiest objective of human striving and existence on earth.

In fact, the very revelation of the Qur'ān is associated with the concept of peace. The 97$^{th}$ chapter known as the Night of Power (*Laylat al-Qadr*) describes the first occurrence of the Qur'ānic revelation as follows:

> We have indeed revealed this (Message) in the Night of Power
>
> And what will cause you to be aware of what the night of power is?
>
> The Night of Power is better than a thousand months.
>
> During which the angels and the Spirit come down by their Lord's permission regarding every matter.
>
> Peace – this is so until the break of dawn!

Embrace of Islam is equated with entering into a peaceful state; Qur'ān 2:108 addresses the believers thus: "O you who believe, enter into peace (*as-silm*) wholeheartedly!" Aṭ-Ṭabarī in his commentary equates *as-silm* with Islam, and mentions that the earlier exegete Mujāhid b. Jabr (d. 104/722) was of the same view. According to aṭ-Ṭabarī, some of the earlier exegetes also understood *as-silm* to mean "peace/peacemaking" (*aṣ-ṣulḥ*) and obedience (*aṭ-ṭāʿa*). The people of the Ḥijāz read the word as *as-salm*, which was equivalent to "peace" and "peaceableness" (*aṣ-ṣulḥ* and *al-musālama* respectively).[90]

Entering into Islam further means entering into the "abode of peace" (Dār as-salām); the Quran (10:25) states: "God summons to the Abode of Peace; He guides whom he pleases to a way that is straight." Accordingly, the speech of righteous believers must be grounded in peace; at a minimum, they invoke on one another peace; even better, they try to outdo one another in exchanging greetings of peace. Qurʾān 4:86 commands: "And when you are greeted with a greeting, greet [in return] with one better than it or [at least] return it [in a like manner]. Indeed, God takes account of everything." One should never churlishly refuse to respond to another's peaceful greeting for whatever reason; the Qur'ān (4:94) admonishes: "Do not say to anyone who offers you peace 'You are not a believer!'" Muslims may invoke peace even on those who harass them for their beliefs and cause them mental anguish. Qurʾān 25:63 praises "the servants of the All-Merciful who walk humbly on earth; and when the foolish jeer at them, they reply, 'Peace!'"

Peace follows the righteous into the next world. When they are admitted into paradise, the Qurʾān (50:34) says they will be greeted by angels who say "Enter in peace! That is the day of eternity!" The believers will hear no offensive speech in paradise, nor any mention of sin, but rather only the saying: "Peace, peace!" (Qurʾān 56:25–26; cf. 56:90–91). And Qurʾān 36:54–56 says that after the Resurrection, "The dwellers in the garden on that day will delight in their situation; they and their spouses will recline on couches in the shade. They will have fruit and whatever they call for. 'Peace!' – the word will reach them from a compassionate Lord."

Peace in the Qurʾān is not just the absence of violence and conflict, which has been called by some a "negative peace." It is rather a holistic concept that is based on two key factors: 1) the recognition of the equal dignity of every human being as God's creation and 2) the guarantee of justice for all on earth. These two key factors allow for peace to be conceptualized as resting on a fundamental restructuring of socio-economic relations and inter-personal relations that are conducive to establishing and nurturing a just and egalitarian society. Such a society can achieve what many call today "positive peace" – that is to say, a lasting and enduring peace because it addresses the root causes of

90 Aṭ-Ṭabarī, *Jāmiʿ al-bayān fi tafsīr al-Qurʾān*, Beirut: Dār al-kutub al-ʿilmiyya, 1997, 2: 335–336.

violence and conflict – inequality and injustice – and seeks to transform society by uprooting them.[91] With regard to the first of the two key factors of positive peace, the Qur'ān (17:70) declares, "Indeed, we have honored every human being," and that every human being has been shaped "in the best of molds" (Qur'ān 95:4), thereby establishing the intrinsic dignity and equal worth of all human beings – regardless of gender, ethnicity, race and social status – by virtue of being human.

With regard to the second factor, a number of Qur'ānic verses establish the necessity of justice as a characteristic of the ideal human community. This is, for example, evident in Qur'ān 57:25 which states: "We sent before Our messengers with clear signs and sent down with them the Book and the Balance (of right and wrong), so that humans may stand forth in justice." Furthermore, Qur'ān 4:135 states "O you who believe! Stand firmly for justice as witnesses before God even if that be against yourselves, your parents, your kin, and whether it be (against) the rich and poor;" while another highly significant verse (Qur'ān 5:8) states, "Let not the hatred of others cause you to incline to wrongdoing and depart from justice. Be just for it is next to piety." These verses clearly establish that justice, according to the Qur'ān, is not only an essential component of peace, but an indispensable one, that allows for human flourishing on earth.

## 7.2 Peace in the Ḥadīth Literature

The ḥadīth literature similarly emphasizes the concepts of non-violence, peace and cordiality in relation to the daily life of the pious Muslim. In the best-known collection of ḥadīth, known as *Ṣaḥīḥ al-Bukhārī* – generally revered by Sunnīs as the most authoritative source after the Qur'ān – the Prophet is quoted as uttering the golden rule, as related by Anas b. Mālik, "No one among you is a believer until they love for others what they love for themselves."[92] The compiler of this collection, al-Bukhārī (d. 256/870), lists another significant ḥadīth narrated by the Companion al-Barā' b. ʿĀzib who related, that Muḥammad had commanded Muslims to habitually visit the sick, attend funerals, quench the thirst of people, come to the aid of the weak, help the oppressed, and spread

91 For this discussion of "negative" vs. "positive peace," see Johann Galtung, *Peace by Peaceful Means. Peace and Conflict, Development and Civilization*, Oslo/London: Sage Publications, 1996.
92 Al-Bukhārī, *Ṣaḥīḥ*, ed. Qāsim ash-Shammāʿī ar-Rifāʿī, Beiurt: Dār al-Qalam, n.d., 1:69, #12.

peace among people.[93] Another Companion ʿAbd Allāh b. ʿUmar narrated, "A man asked the Prophet, peace and blessings be upon him: 'Which Islam is best?'" Muḥammad replied, "The offering of food [to the hungry] and giving greetings of peace to those you know and those you do not know."[94]

The second best-known ḥadīth collection by Muslim b. Ḥajjāj (d. 261/875), also known as *Ṣaḥīḥ*, contains numerous reports that attest to the importance of safeguarding peace and cultivating kindness as characteristics of sincere believers in God. A much-cited ḥadīth is attributed to ʿĀʾisha, the wife of the Prophet, in which she relates that the Prophet addressed her and said, "O ʿĀʾisha, God is most gentle (*ar-rafīq*), and He loves gentleness (*ar-rifq*). Thus He grants to gentleness what He does not grant to harshness/violence (*al-ʿunf*)."[95] According to another ḥadīth cited by Anas b. Mālik, the Prophet counseled his followers: "Do not bear rancor or envy towards one another, and do not oppose one another. Be servants of God and brothers of one another. It is not licit for a Muslim to abandon his brother for more than three [nights]."[96]

A report from Abū Hurayra states that the doors of Paradise are opened every Monday and Thursday and all those who do not ascribe partners to God are forgiven, except for those who harbor a grudge against his brother. The ḥadīth goes on to urge: "Grant them a respite until they reconcile with one another!"[97]

As in the Qurʾān, a number of ḥadīths stress the importance of justice as a corollary to peace and social stability. In a ḥadīth narrated by Abū Hurayra, Muhammad is quoted as saying, "Whoever deals justly with people has carried out an act of charity."[98] In a *ḥadīth qudsī* (sacred statement) narrated by Abū Dharr from Muḥammad, God himself speaks and declares: "I have made tyranny and oppression (*ẓulm*) unlawful for Myself and I have done the same for you. Do not oppress one another!"[99]

Many other ḥadīths in both standard and non-standard collections make the case for peace, non-violence, gentleness, and just behavior to be practiced by the faithful and to exhort others toward it, of which we have presented here a brief but important selection. The concept of peace raises questions about how to establish peace and nurture it when human nature tends to incline

---

**93** Ibid., 7:394, #1109.
**94** Ibid., # 1110.
**95** Muslim, *Ṣaḥīḥ Muslim*, Beirut: Dar Ibn Ḥazm, 1995, 4:1590, #2593.
**96** Ibid., 4:1574, #2559.
**97** Ibid., 4:1577, #2565.
**98** Al-Bukhāri, *Ṣaḥīḥ*, 4:366, #911.
**99** Muslim, *Ṣaḥīḥ*, 4:1583, #2577.

towards conflict and strife. Earlier, we discussed how the Qur'ān advocates peacemaking and the cessation of violence between warring factions, especially when one of the parties is an external, non-Muslim aggressor. But what about circumstances when conflict threatens to erupt between Muslims themselves – what counsel does the Qur'ān provide for effecting reconciliation between erstwhile or potential adversaries within the Muslim community? And even more broadly, what kind of protocol does it establish for effecting enduring peace among people of all good will, especially with members of the two other monotheistic faith communities? It is to this discussion that we turn next.

# 8 The Importance of Peacemaking in the Context of Interpersonal and Communal Relationships

There are a number of verses in the Qur'ān that address the matter of establishing peaceful interpersonal and communal relations. These verses stress the importance of the cultivation of fraternal bonds among fellow believers and the effecting of genuine reconciliation between erstwhile enemies. The centrality of harmonious relationships and the importance of maintaining fraternal concord among believers is articulated very clearly in Qur'ān 49:10: "The believers are brothers, so make peace between your two brothers and be mindful of God, so that you may be given mercy." The next two verses (49: 11–12) stress that believers should be mutually respectful and courteous in their speech and not engage in critical and malicious gossip about one other. Such socially destructive behavior stirs up resentment and ill will and undermines the bonds of fraternity and cordiality that must characterize relations among believers.

The following section looks at additional verses and their exegeses by selected commentators that highlight the importance of the concept of peace in Islamic scripture and tradition, especially in the context of social relationships. These verses are often cited in dialogic contexts with members of the two other Abrahamic faiths as well as in the context of intra-Muslim conversations.

## 8.1 Exegeses of Qur'ān 41:34–35

These verses state "Repel [evil] with what is better, then the one between whom and yourself enmity prevails will become like your friend. But none achieves it [this state of affairs] except for those who are patient and of great fortune."

In the late second/eighth century, Muqātil b. Sulaymān discusses the occasion of revelation for this verse. He says these verses refer to Abū Lahab and his persecution of Muhammad so that the Prophet came to harbor a great dislike for his tormentor. Qur'ān 41:34 counseled him to forgive and pardon Abū Lahab so that his enemy may become transformed into a close friend and ally. The next verse states that only those who show forbearance by suppressing their anger may attain to these good deeds – here good deeds are glossed as forgiveness and clemency (*al-'afū wa-ṣ-ṣafḥ*). And only they will attain to the bounteous blessings of paradise, continues Muqātil, which is the meaning of "a great portion" or "fortune" referred to in this verse (*ḥaẓẓ 'aẓīm*). God had thus exhorted the Prophet to be forbearing in the face of injury and to pray for refuge from Satan in the case of Abū Jahl.[100]

Aṭ-Ṭabarī understands Qur'ān 41:34 to contain a more general injunction for the Prophet to counter the ignorance (*jahl*) of his adversaries with clemency (*ḥilm*) and to disarm those who wish him harm through forgiveness and forbearance. According to Ibn 'Abbas, this verse commands all believers (*mu'minīn*) to be forbearing when provoked to anger and practice magnanimity and forgiveness (*al-ḥilm wa-l-'afū*) in the face of injury. When they do so, God protects them from the machinations of Satan and their enemies relent and become transformed into intimate friends. Other early exegetes, like 'Aṭā' b. Abī Rabāḥ and Mujāhid from the second/eighth century, were of the opinion that the verse required Muslims to respond with greetings of peace (*as-salām*) to those who caused them injury, in explanation of the phrase "with that which is better." Aṭ-Ṭabarī then proceeds to restrict the applicability of this verse by commenting that God had commanded the Prophet to extend such forbearance and forgiveness to his hostile relatives so that they would become his dedicated and compassionate companions. He specifically glosses *al-ḥamīm* as *al-qarīb* or "relative." These verses thus signify both a general injunction to all believers and a more particular meaning rooted in a specific cause of revelation.[101]

As for Qur'ān 41:35, aṭ-Ṭabarī continues, it affirms that only those who show considerable forbearance in the face of wrong-doing and tribulations are able to ward off evil with goodness. "Those who attain to a great portion" are the fortunate ones (*dhū jadd*), according to the early exegete as-Suddī (d. 127/745), while others, like the Companion Ibn 'Abbās, said that *ḥaẓẓ 'aẓīm* specifically referred to paradise. In this connection, Qatāda b. Di'āma is said to have related an anecdote regarding Abū Bakr (d. 13/634), the first caliph, who was

**100** Muqātil, *Tafsir*, 3:743.
**101** Aṭ-Ṭabarī, *Jāmi'*, 11:111.

once cursed by a man in the presence of Muhammad. At first Abū Bakr reacted with forbearance and restraint but then he retorted heatedly, which caused the Prophet to get up and leave. When Abū Bakr pursued him and wished to know the reason for his abrupt departure, Muhammad explained that when the former had exercised self-restraint, an angel supported him, but when he retaliated the angel left and Satan took his place. "And I did not wish to sit in the company of Satan, O Abū Bakr!" remarked the Prophet.[102]

The sixth/twelfth century exegete az-Zamakhsharī illustrates the difference between *ḥasana* and *aḥsan* by offering concrete examples of what constitutes "that which is better" in the face of wrong-doing. An act of goodness (*ḥasana*) in such a situation is to forgive the wrong-doer. To carry out "that which is better" (*wa-llatī hiya aḥsan*) is to respond with an act of goodness or charity to specifically counter or nullify the original injury and thus to go beyond simple forgiveness. Thus, he counsels, if an adversary "were to revile you, praise him; if he were to kill your son, then ransom his son from the hands of his enemy; if you were to carry this out, your inveterate enemy will become transformed into a sincere friend full of good will towards you." Only the people of forbearance and patience (*ahl aṣ-ṣabr*) attain to this and reap goodness (*khayr*) as result. Az-Zamakhsharī cites Ibn ʿAbbās who glossed "with that which is better" as "forbearance when angered, magnanimity in the face of ignorance, and pardon in the face of injury " and understood *al-ḥaẓẓ* in this verse to refer to "reward/ recompense" (*thawāb*) for a good deed. Az-Zamakhsharī notes that the famed successor and pious scholar al-Ḥasan al-Baṣrī however had understood *ḥaẓẓ* to be a reference to "paradise." The specific referent in this verse, concludes az-Zamakhsharī, is Abū Sufyān b. Ḥarb who had been a trenchant foe of the Prophet but then became a sincere ally.[103] Like, aṭ-Ṭabarī, az-Zamakhsharī also points to the possibility of deriving a broad moral injunction from this verse, while bearing in mind the more limited historical context of the verse.

Ar-Rāzī says that the phrase "repel with that which is better" commands the Prophet to ward off the ignorance and foolishness of the ignorant by the best means – by unflaggingly bearing with patience their evil nature, not reacting in anger to their insolence, and by refraining from physical retaliation in return for their injury. It is possible that they will thereby become ashamed of their reprehensible behavior and abandon their repugnant acts, and their

**102** Ibid., 11:111–12.

**103** Az-Zamakhsharī, *Al-Kashshāf ʿan haqāʾiq ghawāmiḍ at-tanzīl wa-ʿuyūn al-aqāwīl fi wujūh at-taʾwīl*, ed. ʿĀdil Aḥmad ʿAbd al-Mawjūd andʿ Alī Muḥammad Muʾawwid, Riyadh: Maktabat al-ʿubaykān, 1998, 5:383.

feelings of animosity will become transformed into affection and good will. Ar-Rāzī cites the well-known lexicographer az-Zajjāj (d. 311/923) who commented, like previous exegetes, that only those who are patient and forbearing in the face of trials and tribulations, suppress their anger, and forego retaliation will attain to the great portion or fortune promised in this verse.[104]

Ibn Kathīr offers a very brief exegesis of these two verses, affirming like most of the previous exegetes, that "repel with what is better" means that one should ward off the harm of the wrong-doer through acts of goodness (*iḥsān*), who is thereby transformed into an affectionate friend and sincere ally. Only those who are patient and practice forbearance – which, according to ʿAlī b. Abī Ṭalḥa, involves restraining one's anger, magnanimity in the face of ignorance, and forgiveness of injury – attain to abundance of happiness in this world and the next, are protected by God from the machinations of Satan, and convert their enemy into a bosom friend.[105]

## 8.2 Exegeses of Qur'ān 3:103

An incipient methodology of effecting reconciliation and fostering peaceful relations among people may be deduced from two other significant Qur'ānic verses: 3:103 and 8:63. These two verses refer to "the joining/bringing together (and therefore reconciliation) of hearts," a process that became known in the extra-Qur'ānic literature as *ta'līf al-qulūb* in Arabic. Muslim exegetes have taken note of these crucial verses and focused on their conciliatory implications in their particular socio-historical contexts as well as in their general applicability, as our discussion below reveals.

Qur'ān 3:103 states,

> And hold firmly to the rope of God all together and do not become divided. And remember God's favor upon you when you were enemies and He joined your hearts together and you became brothers, by his favor. And you were on the brink of a pit of the Fire, and He saved you from it. Thus does God make clear to you his verses so that you may be guided.

Aṭ-Ṭabarī understands the phrase "the rope of God" mentioned in this verse to be the equivalent of "the religion of God which He has commanded you to follow," and, as a consequence of which, believers are vouchsafed "cordiality [among them], unity based upon the word of Truth, and submission to God's

104 Ar-Razi, *Tafsir*, 9:564.
105 Ibn Kathīr, *Tafsir*, 4:103.

commandment."[106] He understands the exhortation "Remember the grace of God" to mean "remember the blessing that He conferred on you of friendship and of being gathered together in Islam." "Remember!" he comments further, "O believers, the bounty of God towards you when you were enemies while polytheists. You killed one another out of tribal partisanship in disobedience to God and his messenger. Then God joined your hearts in Islam (submission [to God]) and made you the brothers of one another after you were enemies and you continue in the bonds of friendship in submission to God."[107]

Aṭ-Ṭabarī provides the occasion of revelation which links the meaning of this verse to the specific situation of two key Medinan tribes before and after the advent of Islam. The enmity referred to in this verse refers to the enmity that existed in the pre-Islamic period between these two tribes known as al-Aws and al-Khazraj as a result of the continuous wars between them. As the pre-Islamic *Ayyām al-ʿArab* ("the Battle-Days of the Arabs") literature informs us, these wars had lasted for about 120 years. Aṭ-Ṭabarī refers to Ibn Isḥāq (d. 151/767), the famous biographer of the Prophet Muhammad, who recalled that these wars took place between al-Aws and al-Khazraj even though they were closely related. According to Ibn Isḥāq, the intensity of their enmity was unprecedented; however, as aṭ-Ṭabarī quotes him, "then God Almighty extinguished that [sc. their enmity] through Islam and brought them together through his messenger, Muhammad, peace and blessings be upon him."[108] Through this verse, aṭ-Ṭabarī affirms, God was reminding the helpers of Muhammad in Medina (*anṣār*) of the misery and wretchedness that had afflicted them on account of their mutual hostility and fear which had led to the shedding of each other's blood. The verse is a further reminder of how submission to God and the guidance of the Prophet finally led to reconciliation and security and facilitated the development of the bonds of friendship and brotherhood between them.[109]

Aṭ-Ṭabarī concludes this section by commenting on the last part of Qur'ān 3:103 ("and you became brothers by His grace"). This means, he remarks,

> Because God, exalted is He, joined your hearts by means of Islam and the word of Truth. Through cooperation in aiding the people of faith and rallying against unbelievers who opposed you – so that there remained no malice and no enmity between you – you became sincere, truthful brothers.[110]

**106** Aṭ-Ṭabarī, *Jāmiʿ*, 3:378.
**107** Ibid.
**108** Ibid., 7:78.
**109** Ibid.
**110** Ibid., 7:84.

The seventh/thirteenth century Andalusian exegete al-Qurṭubī understands the "rope of God" mentioned at the beginning of Qur'ān 3:103 to refer to the Qur'ān as well as to communal life. Al-Qurṭubī remarks that the meanings are intertwined and related, for God has enjoined sociability and forbidden separation from the community, for "separation is [equivalent to] destruction and communal affiliation is [equivalent to] salvation."[111]

The prohibition "do not become disunited" warns against the divisiveness that can afflict religious communities in the practice of their religion and exhorts believers to remain brothers in "the religion of God" by not blindly following every whim and fancy. Al-Qurṭubī goes on to clarify, however, that the verse does not rule out differences with regard to secondary matters of religion (Ar. *al-furūʿ*). If such differences do not lead to dissension and arise in the course of extrapolating legal rulings from and explicating the religious law, then they are not harmful in themselves. The Companions of the Prophet often differed with one another in the derivation of legal rulings in new circumstances, but in spite of that, remained amicable towards one another. Al-Qurṭubī cites the well-known *ḥadīth* "The difference of my community is a mercy." The kind of difference God forbids is the kind that leads to corruption and sectarianism.[112]

In his commentary on Qur'ān 3:103, the popular eighth/fourteenth century exegete Ibn Kathīr (d. 774/1352) similarly stresses the "joining of hearts" that can come about even among the bitterest of enemies when humans are open to the transformative power of God's love in their relations with one another. He points to the Aws and Khazraj tribes who were prone to much fighting in the pre-Islamic period (*jāhiliyya*), and harbored virulent hatred towards one another. All that was to change, however, "when God brought forth Islam, and those who entered it, entered it and became brothers, loving one another through the majesty of God, cooperating with one another in a spirit of piety and God-consciousness."[113] Ibn Kathīr further points to the danger inherent in relaxing one's guard against potential fractiousness and thus allowing seditious elements to stir up enmity. He relates an anecdote in exegesis of Qur'ān 3:103, according to which an individual of ill will, who was displeased to see the concord and amity reigning among the Aws and Khazraj tribes in Medina, dispatched one of his cohort to sit among them and revive the memory of their prolonged wars during the *jāhiliyya*. The man did as he was told and before long, he had succeeded in inflaming the passions of the

---

**111** Al-Qurṭubī, *Al-Jāmiʿ li-aḥkām al-qur'ān*, Cairo: Dār wa-maṭābīʿ ash-shaʿb, n.d., 2:1401.
**112** Ibid., 2:1402ff.
**113** Ibn Kathīr, *Tafsīr*, 1:368.

Aws and Khazraj to the extent that they started chanting their battle slogans from the pre-Islamic period and reaching for their weapons. News of this reached the Prophet and he hurriedly approached them and began to calm them down. He asked, "Are you harking back to the pre-Islamic period while I am among you?" Then he recited to them Qur'ān 3:103, at which they were filled with remorse. "They began to make up with one another, and embraced one another after having cast away their weapons."[114]

This report cogently demonstrates the recuperative power of the remembrance of God and his limitless grace towards humans, which effaces the memory of past wrongs and allows for forgiveness and reconciliation to occur among the bitterest of enemies. But, it also warns, such memories can be resurrected by those malignantly inclined to sow dissension. The constant invocation of God in gratitude for his immeasurable benevolence towards humankind is, however, a potent shield against the incitements of troublemakers and helps preserve the unity of the believers.

## 8.3 Exegeses of Qur'ān 8:63

This verse states: "And He [God] joined their hearts together. If you had spent all that is in the earth, you could not have joined (reconciled) their hearts; but God brought them together. Indeed, He is Mighty and Wise."

The exegeses of this verse are very similar to those for Qur'ān 3:103. Most commentators, like aṭ-Ṭabarī, al-Qurṭubī, and Ibn Kathīr, understand this verse to be a reference to the Aws and Khazraj tribes and the healing of their bitter past and mutual forgiveness after they became believers, as well as to the brotherly love that was created between the Meccan Emigrants and the Medinan Helpers in the post-*hijra* period. The verse is furthermore understood to stress that only God when he enters our hearts can effect this kind of transformation. Thus, aṭ-Ṭabarī comments that this verse addresses the Prophet and reminds him, "Were you to spend, O Muhammad, all that is on earth of gold, money, and worldly goods, you would not be able to reconcile their hearts with all your strength, for it is God Who joins them in guidance!"[115]

Aṭ-Ṭabarī refers to the earlier Medinan exegete Mujāhid b. Jabr (d. 104/720) who extrapolated a general mandate for reconciliation and peacemaking from

**114** Ibid.
**115** Aṭ-Ṭabarī, *Jāmiʿ*, 6:280.

this verse. Thus, Mujāhid says that the statement "Were you to spend all that is on earth you would not be able to reconcile their hearts," means that when two Muslims meet and shake hands, their sins are forgiven. A variant exegesis attributed to Mujāhid offers further clarification of this somewhat elliptical comment. According to this variant, a man named ʿAbda b. Abī Lubāba related that he met Mujāhid and the latter took his hand in his own and said, "If you should see two individuals who harbor love for God and one of them takes the hand of the other and smiles at him, their sins drop off them just as the leaves drop from the tree." ʿAbda told Mujāhid, "But indeed that is easy." Mujāhid remarked, "Do not say that, for indeed God has stated, 'Were you to spend all that is on earth you would not be able to reconcile their hearts.'"[116]

The two reports taken together convey Mujāhid's conviction that sincere faith in God results in genuine bonds of friendship and good will among believers, expressed outwardly in gestures of friendship towards one another, such as by shaking hands and exchanging smiles. But simply going through such motions does not automatically create a sense of bonhomie, unless they are firmly embedded in faith and love for God – this latter being a much harder task, as pointed out by Mujāhid – and can only be effected by God himself. Once firmly implanted in one's heart, love for God translates into love for one's fellow beings.

# 9 The Modern Period

In the modern period, Muslim scholars have often pushed back against the frequently negative depictions of Islam and its supposed proclivity for violence by many Western missionaries and Orientalist scholars, particularly during the period of European colonization of a broad swath of the Muslim world during the late eighteenth, nineteenth and the early part of the twentieth centuries. Below we focus on one such prominent scholar.

## 9.1 Muḥammad ʿAbduh

The nineteenth century Muslim reformist and scholar Muḥammad ʿAbduh (d. 1905) who lived during the period of British colonization of his native

116 Ibid., 6:280–81.

Egypt, often criticized Orientalist characterizations of *jihād* as a relentless militant activity. ʿAbduh was also critical of a number of medieval Muslim exegetes who invoked the concept of abrogation (*naskh*) to articulate a conception of the military *jihād* as offensive warfare that could be waged against non-Muslims. Thus, ʿAbduh rejects the interpretation that the so-called "sword verse" (Qurʾān 9:5) had abrogated the more numerous verses in the Qurʾān which call for forgiveness and peaceful relations with non-Muslims. Citing the views of as-Suyūṭī (d. 911/1505), ʿAbduh argues that in the specific historical situation with which the verse is concerned – with its internal reference to the passage of the four sacred months and to the pagan Meccans – other verses in the Qurʾān advocating forgiveness and non-violence were not abrogated but rather in [temporary] abeyance or suspension (*laysa naskhan bal huwa min qism al-munsaʾ*). *Naskh* implies the abrogation of a command, which is not the case here. Rather the command contained in Qurʾān 9:5 was in response to a specific situation at a specific time in order to achieve a specific objective and has no effect on the injunction contained in, for example, Qurʾān 2:109, which states, "Pardon and forgive until God brings about His command," which is in regard to a different set of circumstances and objectives.[117]

ʿAbduh is critical of those who would see the injunction contained in Qurʾān 9:5 with its clear reference to Arab polytheists applicable in any way to non-Arab polytheists or to the People of the Book. The latter are referred to very differently in the Qurʾān, as in Qurʾān 5:82,[118] and in *ḥadīth*s, such as the one which counsels leaving the Ethiopians (as well as Turks) alone as long as they leave the Muslims alone. He bemoans the fact that if jurists had not read these verses and *ḥadīth*s "from behind the veil of their juridical schools" then they would not have so egregiously missed the fundamental point made throughout the Qurʾān and in sound *ḥadīth*s that "the security to be obtained through fighting the Arab polytheists according to these verses is contingent upon their initiating attacks against Muslims and violating their treaties . . . "[119] ʿAbduh goes on to point out that the very next verse Qurʾān 9:6 offers protection and safe conduct to those among the polytheists who wish to listen to the Qurʾān.[120] The implication is clear – polytheists and non-Muslims in general who do not wish

**117** Muḥammad ʿAbduh and Rashīd Riḍā, *Tafsīr al-manār*, Cairo: Maṭbaʿat al-manār, 1931, 10: 161–62.

**118** This verse states, "You will find the closest in affection to those who believe are those who say we are Christians."

**119** Riḍā, *Tafsīr al-manār*, 10:162–63.

**120** Ibid., 10:171–75.

Muslims harm and display no aggression towards them are to be left alone and allowed to continue in their ways of life.

ʿAbduh identifies three different types of *jihād*: 1) struggle against the external enemy (*mujāhadat al-ʿaduw aẓ-ẓāhir*); 2) struggle against the devil (*mujāhadat ash-shayṭān*); and 3) struggle against the soul (*mujāhadat an-nafs*). All three types are included in the following Qurʾānic injunctions: "Strive in regard to God as is His due" (Qurʾān 22:78); "Strive with your wealth and selves in the path of God" (Qurʾān 9:41); and "Those who believed, emigrated, and strove with their wealth and their selves in the path of God" (Qurʾān 9:72). Two *ḥadīth*s furthermore attest to the manner of carrying out *jihād* by the hand and the tongue: one in which Muhammad says, "Struggle against your passions (*ahwāʾakum*) as you struggle against your enemies;" and the other in which he says, "Strive against the unbelievers with your hands and your tongues." The latter *ḥadīth*, continues ʿAbduh, stresses the primacy of *jihād* of the tongue – that is, of attesting to the truth by means of amassing evidence and compelling arguments.[121]

The above proof-texts and others beside them belie the arguments made by Orientalist scholars and those who follow them that *jihād* is reducible to fighting against non-Muslims in order to forcibly effect their conversion. ʿAbduh points to Qurʾān 2:256 ("There is no compulsion in religion") and other verses which allow fighting only against those who initiate fighting and which command Muslims to incline to peace when the adversary inclines to peace as proof-texts – all of them establish the falsity of imputing such a reductive meaning to *jihād*.[122] Wars fought for material gain and for the shedding of blood, as was common among ancient kings, or for revenge and out of religious animus, as in the case of the Crusades, or for the purpose of confiscating the possessions of the weak and demeaning human beings, as evident in the European colonial wars of his time, are all forbidden by Islamic law, he says.[123]

With regard to Qurʾān 3:103, ʿAbduh, like a number of his pre-modern predecessors, understands the verse to be a reference to the reconciliation of the Aws and Khazraj tribes of Medina after their submission to God, putting an end to their bitter past of chronic hostility. He further understands this verse to

**121** Ibid., 10:279.

**122** Many of these points are also made strenuously by other modern Muslim scholars, such as Abū Zahra, *al-ʿAlāqāt ad-dawliyya fī l-islām*, Cairo: Maṭbaʿat al-azhar, 1964, 47:52; Subhi Mahmassani, "The Principles of International Law in the Light of Islamic Doctrine," *Recueil des Cours* 117 (1966): 249–79; Wahba az-Zuḥaylī, *Āthār al-ḥarb fī l-fiqh al-islāmī: dirāsa muqārana*, Damascus: Dār al-fikr, 1982, 503, and others.

**123** ʿAbduh and Riḍā, *Tafsīr al-manār*, 10:280.

contain a strong denunciation of the tribalism of the pre-Islamic period, termed in Arabic *al-ʿaṣabiyya*. He marshals as proof-text the *ḥadīth* in which the Prophet declares, "One who invokes tribalism is not one of us." ʿAbduh sees this pre-Islamic tribalism resurgent in the nationalisms of his own time which create dangerous divisions among people. ʿAbduh asserts that the true advancement of a nation lies in uniting all its citizens through their devotion to God, which ensures the well-being and welfare of all people, regardless of their religion or ethnicity.[124] It is through "holding fast to God's rope" that one may successfully resist divisiveness and sectarianism which leads to the shedding of blood, as happened in the pre-Islamic past, and thereby achieve genuine reconciliation among people.

With regard to Qur'ān 8:63, ʿAbduh understands this verse to apply primarily to the Meccan Muslims, who became brothers of the Medinan Muslims in faith, despite differences in social status and worldly rank. He underscores this dramatic transformation in the following way: "As for the Muhājirūn, reconciliation (*ta'līf*) occurred among their rich and the poor, their masters and their clients, their nobility and their common people, in spite of the arrogance of the Jāhiliyya that had previously existed among them."[125] It was this concord among them that allowed them to endure the enmity of their fellow tribesmen and relatives for the sake of God. None of this could have been achieved by means of all the wealth and enticements of the world.

ʿAbduh then goes on to point to the centrality of love in human relationships, which has been, he says, asserted by wise people through the ages. These sages agree that "Love is the greatest of all bonds among humans and the most potent inducement to happiness is human social life and its refinement."[126] They further concur that in the absence of love, nothing else can take its place in repelling evil, while the proper functioning of society is contingent on the dispensation of justice. While love has been considered to be instinctual and not a matter of choice and justice regarded as an act of deliberation, Islam made love a virtue and adherence to justice an obligatory duty. Justice in particular was the due of all who reside in the Islamic state, with no distinction to be made between the Muslim and non-Muslim, pious and impious, rich and poor, etc.[127]

In this important exegesis, ʿAbduh goes further than his pre-modern predecessors and extends the concept of reconciliation based on love and justice to

**124** Ibid., 4:21.
**125** Ibid., 10:70–71.
**126** Ibid., 10:71.
**127** Ibid.

all human beings, regardless of their religious affiliation (or lack thereof). He argues that out of love for the Creator and adherence to justice the individual and the state must treat everyone even-handedly.

## 9.2 *Jihād* as Peaceful Activism

There are several contemporary scholars who have focused in their written works on the peaceful activism they understand to be the predominant meaning of *jihād*. A number of such scholars and thinkers typically emphasize the virtue of patient forbearance as the most important aspect of *jihād*, and therefore of non-violent resistance to wrong-doing. This modern emphasis on non-violent public activism as the best manifestation of *jihād* has been espoused by a number of well-known and less well-known figures. One of the more prominent names from the twentieth century is that of the Pashtun leader Syed ʿAbd al-Ghaffār Khān (d. 1988). He organized a peaceful resistance movement called the Khudai Khidmatgars ("the Servants of God") against the British colonizers of India, arguing that Muslims should adopt non-violence against oppression on the basis of their own scriptural directives and historical praxis of the early Muslims which emphasized patience (*ṣabr*).[128] For a closer study of this "school" of non-violence based on published materials, I am presenting the thinking of Jawdat Saʿīd and Wahiduddin Khan, who are among the best-known contemporary writers on this topic. Some of the key points of their arguments in favor of non-violence are discussed below.

### 9.2.1 Jawdat Saʿīd

Jawdat Saʿīd (b. 1931) is a well-known Syrian writer and thinker known for his pacifist views, derived from his reading of the Qurʾān, particularly of the story of Adam's two sons, as elaborated below. He obtained a degree in Arabic language from al-Azhar University and eventually settled in Bir Ajam in the Golan Heights, where he lives in the ancestral family house until today. In the English

---

**128** For a detailed study of Abdul Ghaffar Khan's movement, see Robert C. Johansen, "Radical Islam and Nonviolence. A Case Study of Religious Empowerment and Constraint among Pashtuns," *Journal of Peace Research* 34 (1997): 53–71. For a monograph-length study, see Eknath Easwaran, *A Man to Match His Mountains. Badshah Khan Nonviolent Soldier of Islam*, Petaluma, CA: Nilgiri Press, 1984.

translation of his work titled *Non-Violence: The Basis of Settling Disputes in Islam*,[129] Saʿīd grounds his non-violent understanding of *jihād*, glossed as the struggle to resist wrong-doing, in his reading of the Qurʾānic verses (5:27–31) which give an account of the violent altercation between Adam's two sons. These verses state:

> And recite to them the story of Adam's two sons, in truth, when they both offered a sacrifice [to God], and it was accepted from one of them but was not accepted from the other. Said [the latter], 'I will surely kill you. Said [the former], 'Indeed, God only accepts from those who are righteous [who fear Him]. If you should raise your hand against me to kill me – I shall not raise my hand against you to kill you. Indeed, I fear God, Lord of the worlds. Indeed, I want you to obtain [thereby] my sin and your sin, so you will be among the companions of the Fire. And that is the recompense of wrongdoers.' And his soul permitted him to murder his brother, so he killed him and became among the losers. Then God sent a crow searching [i.e., scratching] in the ground to show him how to hide the private parts of his brother's body. He said, 'O woe to me! Have I failed to be like this crow and hide the private parts of my brother's body?' And he became of the regretful.[130]

Among the relevant ethical and moral imperatives that Saʿīd derives from these verses are a) that a Muslim should not call for murder, assassination, and/or any provocative acts that may lead to the commission of such crimes; b) that a Muslim should not present his opinion to others by force or yield to others out of fear of any such force; and c) that a Muslim in his/her pursuit to spread the word of God "must not diverge from the true path which was set forth by the prophets from beginning to end."[131] The third inference indicates Saʿīd's understanding of *jihād* as an essentially non-violent enterprise undertaken by Muslims for the purpose of bearing witness to the truth and justice of their faith and to propagate it – in other words – to carry out *daʿwā*, which he defines as "an act of calling . . . to Islam."[132]

Muslims, continues Saʿīd, are primarily entrusted with speaking "the words of truth under any condition."[133] In this context, he refers to the *ḥadīth* in which Muhammad affirms that the best *jihād* is speaking a word of truth to a tyrannical ruler. Our author further suggests that while being a witness to truth in this manner, a Muslim may not resort to violence, even apparently in

**129** Translated by Munther A. Absī and H. Hilwānī, Damascus: Dār al-fikr, 2002 from the original Arabic.

**130** Translation in Saʿīd, *Non-Violence*, 27.

**131** Ibid., 35.

**132** Ibid., 34.

**133** Ibid., 37.

self-defense. He refers to the *ḥadīth* in which Saʿd b. Abī Waqqās asked the Prophet what he should do if someone were to come into his house and "stretches his hand to kill me?" The answer was, "Be like Adam's [first] son;" and then Muhammad recited Qur'ān 5:27–31.[134]

But what about the combative *jihād* which the Qur'ān clearly permits under certain conditions? Saʿīd does not deny that these verses exist but states that their commands are not applicable in the absence of a properly formed Islamic community, which is currently the situation in which Muslims live. A properly formed Islamic community is one in which truth and justice reign, inhabited by Muslims "who call for the construction of the Islamic society, its reformation or protecting it against the elements of corruption." They are furthermore those

"who have enough courage to declare their creed and everything they believe in, and who are openly denouncing what they believe to be wrong in a clear way (thus reaching (sic) the distinct propagation of Islam. . . . They are the kind of people who, for their cause, persevere patiently with the oppression of others when they are subjected to torture and persecution."[135]

Such patient, non-violent activism in the face of oppression and injustice and in the absence of the properly constituted Islamic community is the only form of *jihād* that can be carried out by Muslims today, asserts Saʿīd. Such non-violent activism is in emulation of all the prophets mentioned in the Qur'ān who patiently endured the harm visited upon them by their own people on account of their preaching the truth. One of the examples our author highlights is that of Moses arguing calmly and peacefully before the Pharaoh in defense of the truth that he had been called to preach. In contrast, the Pharaoh resorted to aggression, as tyrannical rulers are apt to do, in order to protect their political dominion.[136] Believers should not resort to violent overthrow of despotic governments, counsels Saʿīd – for then they would be following in the footsteps of the Pharaoh by adopting violent methods. Like Moses and all the other prophets, they should attempt instead to bring about a peaceful resolution of conflict through the clear and fearless proclamation of the truth.[137]

It should be noted that Saʿīd does not state that fighting is always categorically prohibited; he recognizes *jihād* "as an ongoing process on condition that a Muslim must know exactly when to resort to armed struggle."[138] "Executing

**134** Ibid., 28–29.
**135** Ibid., 78.
**136** Ibid., 40–57.
**137** Ibid., 37–40.
**138** Ibid., 39.

laws," he says, "and carrying out Jihād must only be done by individuals who are qualified for such an important task."[139] The improper and excessive recourse to the combative *jihād* and cynical manipulation of it by unscrupulous people have "caused more harm to Muslims than any other malpractice."[140] Muslims are primarily charged today with preaching the message of God and reforming humans, which can never be accomplished by force as stated in the verse "Let there be no compulsion in religion" (Qur'ān 2:256).[141] Sa'īd calls those who advocate unconditional violence in the name of Islam "preachers of terrorism" whose vicious ideology "must be quelled with any possible means."[142] Evil cannot be erased by violence, however; evil can only be eradicated by the establishment of justice, and justice is served by the best form of *jihād* – the proclamation of truth.[143] Sa'īd stresses that such truth should be presented on the basis of reason and should conform to Qur'ānic evidential standards, as stated in Qur'ān 2:111, "Say, 'Produce your proof, if you should be truthful.'"[144] The Qur'ānic exhortation to acquire knowledge through reflection and travel ("Travel through the land and observe how He began creation;" Qur'ān 29:20) requires fundamental cognitive and spiritual changes among Muslims today, which are the prelude to broader social transformations (cf. Qur'ān 8:53).[145] When humans are able to comprehend God's signs more fully, then they will begin to apprehend the root causes of their behavioral problems and proceed to solve them. "At that point the society will recover from all the causes which make people turn against each other, the same as those who recover from diseases befalling their bodies."[146] Violence is a disease which afflicts us all and threatens to engulf us unless a comprehensive revolution changes human attitudes, "especially since we are still bound within the phase of the belief in the accusations which the Angels launched against Adam, as being a creature who promotes destruction and corruption."[147] This is the message and mission that Sa'īd wishes to convey to the youth in particular so that they may be able to bring about these necessary peaceful transformations.

**139** Ibid., 122.
**140** Ibid., 40.
**141** Ibid., 62.
**142** Ibid., 74.
**143** Ibid., 77–79.
**144** Ibid., 111.
**145** Ibid., 114.
**146** Ibid., 124.
**147** Ibid., 124–25.

### 9.2.2 Wahiduddin Khan

Wahiduddin Khan, born in 1925, is a contemporary Indian scholar of Islam who is the president of the Islamic Centre in New Delhi, India. For fifteen years he was a member of the Jama'at-i Islami founded by Mawdudi in 1941 but broke with the latter because of fundamental disagreements concerning the relation between Islam and politics. Khan emphasized, unlike Mawdudi, that *tawḥid* and peaceful submission to God was at the heart of all things Islamic and not political and economic reform.[148]

In his book *The True Jihād: The Concept of Peace, Tolerance and Non-Violence*[149] written in the aftermath of September 11, Khan stresses that the main purpose of Islam was the peaceful propagation of the faith (*daʿwā*) and that political and social reform were at best secondary concerns which would inevitably result from the spiritual reformation of Muslims. He begins this short treatise by pointing to Qur'ān 22:78 which exhorts the believer to "strive for the cause of Allah as it behooves you to strive for it." *Jihād* derived from the Arabic root *j-h-d* points to this earnest struggle for the sake of God, a term which eventually came to be applied to the early battles in Islam as well, since they were part of this overall struggle. Strictly speaking, the term for fighting is *qitāl*, and not *jihād* per se. On the basis of the *Musnad* of Aḥmad b. Ḥanbal, he identifies the *mujāhid* as "one who struggles with himself for the sake of God;" as "one who exerts himself for the cause of God;" and as "one who struggles with his self in submission to the will of God." *Jihād* is therefore essentially a peaceful struggle against one's ego and against wrong-doing in general.[150]

Khan proceeds to establish the peaceful essence of *jihād* by invoking the following proof-texts. He refers to Qur'ān 25:52 ("Do not yield to the unbelievers, but fight them strenuously with it [the Qur'ān]"), which establishes that *jihād* is essentially a peaceful, non-violent struggle to establish the truth since "no military activity is referred to in this verse." A *ḥadīth* narrated by ʿĀ'isha, recorded by al-Bukhārī, quotes the Prophet as expressing a preference for the easier of any two options. Since war is a hardship, this *ḥadīth* encodes the superiority of the peaceful struggle for truth. The Prophet's biography reveals that he never initiated hostilities and that he went to great lengths to avoid it. Examples from his life which support this interpretation are as

**148** Cf. the article by Irfan A. Omar, "Islam and the Other. The Ideal Vision of Mawlana Wahiduddin Khan" *Journal of Ecumenical Studies* 36 (1999): 423–39.
**149** Published by Goodword Books, New Delhi, 2002.
**150** Khan, *True Jihad*, 13–16.

follows: 1) In the Meccan period, Muhammad was primarily concerned with challenging polytheism through peaceful, verbal means; 2) Even when during the thirteen year Meccan period the Quraysh became his arch-enemy and prominent members of the tribe conspired to kill him, he avoided any physical confrontation and resorted instead to migration to Medina at the end; 3) the battle of the Trench is a stellar example of avoiding unnecessary violence; as is 4) the Treaty of al-Hudaybiyya which the Prophet signed with the pagan Meccans in order to avoid the shedding of blood; and 5) the peaceful conquest of Mecca at a time when the Muslims were militarily strong testify to the preference for non-violent methods over violent ones to promote truth and justice. These examples provide testimony, states Khan, that "the position of peace in Islam is sacrosanct, while war in Islam is allowed only in exceptional cases when it cannot be avoided."[151]

Muslim advocacy of the principle of non-violence today recognizes "that the commands of the shariah change according to altered situations."[152] In the premodern period, war was a way of life; now we are able to imagine and implement peaceful strategies for conflict resolution. Khan scoffs at "the *jihād* movements" of the contemporary period for their glorification of violence; in these changed circumstances, "launching out on a violent course of action is not only unnecessary, but also un-Islamic."[153] A movement, he says derisively, cannot be deemed a *jihād* "just because its leaders describe it as such."[154] A properly constituted *jihād* must fulfill the essential conditions decreed by Islamic law. The combative *jihād* which is essentially *qitāl* (glossed as "armed struggle") is an activity relating wholly to the state and cannot be placed in the same category as acts of worship, such as prayer and fasting. There is no room, he emphasizes, for non-state warfare, for war, and it must be defensive war, may be declared only by the ruling government. Non-combatants may not be targeted. On this basis, Khan sternly condemns the perpetrators of the September 11 attacks. He also proscribes the carrying out of suicide bombings which he declares to be a complete departure from Islamic norms and religiously-sanctioned practices.[155] Khan comments, "According to Islam we can become martyrs, but we cannot court a martyr's death deliberately."[156]

**151** Ibid., 16–23.
**152** Ibid., 25.
**153** Ibid., 26.
**154** Ibid., 27.
**155** Ibid., 23–38.
**156** Ibid., 39.

The Qur'ān makes a fundamental difference between "the enemy" and "the aggressor," continues Khan. Believers have not been granted the right to wage unprovoked wars against their enemies; the Qur'ān actually commands them to wage peace against them instead. How? Qur'ān 41:33–34 instructs them: "And good and evil deeds are not alike. Repel evil with good. And he who is your enemy will become your dearest friend."[157] Khan discerns in these verses a clear Qur'ānic mandate for "turning one's enemy into a friend through peaceful means, instead of declaring him an enemy and then waging war against him." Muslims may resort to fighting only if the enemy attacks them first and only when all efforts at reconciliation and peaceful resolution of the conflict have failed. Muslims are clearly forbidden to initiate wars except in response to a prior act of violent aggression, as in Qur'ān 22:38 ("Permission to take up arms is hereby given to those who are attacked because they have been wronged") and in Qur'ān 9:13 ("They were the first to attack you").[158]

Commands to fight in the Qur'ān are to be understood as "specific to certain circumstances" and "were not meant to be valid for all time to come."[159] Islam is fundamentally a religion which teaches non-violence, he asserts. The Qur'ān states that God does not love *fasād*, which Khan glosses as "violence." Qur'ān 2:205 clearly indicates, he comments, that "*fasād* is that action which results in disruption of the social system, causing huge losses in terms of lives and property." God loves non-violence; and He promises in Qur'ān 16:5 that "Those who seek to please God will be guided by him to 'the paths of peace.'" As a consequence of this high valorization of non-violence, the Qur'ān eulogizes patience (*ṣabr*) as a human virtue, promising reward for it that is beyond measure (Qur'ān 39:10). *Ṣabr* is the equivalent of non-violence as understood in the modern period. The absolute higher valuation of non-violence over violence is indicated in a *ḥadīth* in which the Prophet remarks, "God grants to *rifq* (gentleness) what he does not grant to *'unf* (violence)."[160]

Non-violent activism is particularly relevant for Muslims in the contemporary period and is the most important aspect of *jihād* for them today, affirms Khan. Peaceful interactions between Muslims and non-Muslims will allow for serious dialogue to emerge between them and expose Muslims to the kind of intellectual stimulation they are badly in need of "if they are to tread the path

**157** Ibid., 39–40.
**158** Ibid.
**159** Ibid., 44–45.
**160** Ibid., 46–48. For the importance of *ṣabr* as a basic principle of non-violence and peace-building, see also Mohammed Abu-Nimer, *Nonviolence and Peace Building in Islam. Theory and Practice*, Gainesville: University Press of Florida, 2003, 71–73.

of progress."[161] Adopting the path of non-violence, continues our author, will be tantamount to "reviving the sunnah of Hudaybiyya;" an event the Qur'ān (48:26) had referred to as "a clear victory."[162] Ideally, peace should be accompanied by justice. But so strong is the imperative towards non-violence in Islam, asserts Khan, that one may settle for peace first even if it falls short of justice, as was exemplified by the Prophet's agreement to the terms of al-Hudaybiyya, which were unfavorable towards Muslims. This acceptance of a lopsided peace treaty did however lead to the establishment of justice and made unnecessary the waging of war to attain it. He reminds that "God calls to the Abode of Peace" (Qur'ān 10:25) and there is no other way to realize God's will.[163]

# 10 Conclusion

Our exploration of the historical trajectory of *jihād* in the context of war and peacemaking leads us to the following conclusions. First, this study documents the multiple and contested meanings of *jihād* that are prevalent in different genres of sources consulted here – Qur'ān, *ḥadīth*, legal and ethical/edifying literature – and challenges a monolithic, reductive understanding of the term. Second, it establishes the defensive and limited nature of legitimate war in the Qur'ān as stressed particularly by exegetes, ethicists, and moral theologians. In the Qur'ān, peace is the default situation; war can be waged only as a last resort when other peaceful means of resolving conflict have been exhausted and Muslims have been attacked by the enemy. *Jihād* in the Qur'ān is therefore most categorically not holy war, as it is often (mis)translated into English (and its equivalent in other Western languages). Holy war is aggressive war waged in the name of God to effect the forcible conversion of non-believers and is often a "total, no-holds barred war" intended to annihilate the enemy.[164] Both of these objectives are doctrinally unacceptable in

**161** Khan, *True Jihad*, 94.
**162** Ibid., 95.
**163** Ibid., 105–108.
**164** Roland Bainton's definition of holy war in the context of the Crusades is generally accepted; he described the Crusades as "a holy war fought under the auspices of the church or some inspired religious leader, not on behalf of justice conceived in terms of life and property, but on behalf of an ideal, the Christian faith"; see his *Christian Attitudes toward War and Peace*, Nashville, Tenn.: Abingdon, 1986, 14.

Islam. Third, it contextualizes the legal positions that legitimized offensive military activity as contingent responses to specific political circumstances, which cannot be deemed to be normatively binding for Muslims for all times and for all places. This is a position that emerges very clearly in the writings of modern Muslim theologians like Muḥammad ʿAbduh and others.

The Qur'ān also strongly advocates for peaceful interpersonal and intercommunal relationships among Muslims and between Muslims and non-Muslims, especially the adherents of the two other monotheistic faith communities. The verses discussed above concerning this aspect of peacemaking (Qur'ān 3:103; 8:63) and their exegeses by some of the most prominent Muslim commentators through time have much to tell us today about faith-based resolution of conflictual situations and harmonious coexistence with others. These verses locate both love and animosity within the human heart; which of the two gets the upper hand within it is contingent upon certain choices of the individual. The individual, on the one hand, may choose to believe in and submit to God, thereby cleansing his or her heart of resistance to God's will and allowing one's heart to be flooded with love for God and, consequently, for God's creation. On the other hand, one can reject faith in God and harden one's heart against other human beings and thus allow oneself to be swept away by worldly needs and the desire for dominance. The two diametrically opposed states are exemplified by the Medinan tribes of Aws and Khazraj, who were intractable enemies before the advent of Islam. But, once faith entered their hearts through their submission to God, it was God, as the exegetes remind us, who transformed their inward state from one of animosity to comradely love and peaceful reconciliation.

More recently, several influential Muslim thinkers have resuscitated the concept of *ṣabr* as the most important element of *jihād* that can be deployed as a guiding principle for promoting non-violence and peacemaking in the modern world. This development constitutes a robust recognition of the centrality of this concept in the lexicon of contemporary Islam – both from an ethical and praxis-based perspective. Through a close reading of scripture and the historical contextualization of later literary productions which chart the storied history of *jihād*, we can assert that fundamental Islamic perspectives on peace and war have much to contribute to contemporary global discussions concerning violence and conflict resolution. Peacemaking within Islam is scripturally mandated and woven into the religion's very foundation – this is a message that is timely and urgent in the divisive times that we live in.

# Bibliography

ʿAbduh, Muḥammad/Riḍā, Rashīd, *Tafsīr al-manār*, Cairo: Maṭbaʿat al-manār, 1931.
Abu-Nimer, Mohammed, *Nonviolence and Peace Building in Islam. Theory and Practice*, Gainesville: University Press of Florida, 2003.
Abū Zahra, *Al-ʿAlaqāt al-dawliyya fī ʾl-islām*, Cairo: Maṭbaʿat al-azhar, 1964.
Afsaruddin, Asma, *Striving in the Path of God. Jihad and Martyrdom in Islamic Thought*, Oxford: Oxford University Press, 2013.
Afsaruddin, Asma, *Excellence and Precedence. Medieval Islamic Discourse on Legitimate Leadership*, Leiden: Brill, 2002.
Afsaruddin, Asma, "Competing Perspectives on Jihad and Martyrdom in Early Islamic Sources," in: Brian Wicker (ed.), *Witnesses to Faith? Martyrdom in Christianity and Islam*, Aldershot, U.K.: Ashgate Publishing, 2006.
Bainton, Ronald, *Christian Attitudes toward War and Peace*, Nashville, Tenn.: Abingdon, 1986.
Bostom, Andrew, *The Legacy of Jihad. Islamic Holy War and the Fate of Non-Muslims*, New York: Prometheus Books, 2005.
Ceadel, Martin, *Pacifism in Britain 1914–1945. The Defining of a Faith*. Oxford: Clarendon Press, 1980.
Ceadel, Martin., *Thinking about Peace and War*, Oxford: Oxford University Press, 1987.
Cook, David, *Understanding Jihad*, Berkeley: University of California Press, 2009.
Al-Dhahabī, Shams al-Dīn, *Mīzān al-iʿtidāl fi naqd al-rijāl*, ed. Badr al-Dīn al-Naʿsānī, Cairo: al-Khanjī, 1907.
Easwaran, Eknath, *A Man to Match His Mountains. Badshah Khan Nonviolent Soldier of Islam*. Petaluma, CA: Nilgiri Press, 1984.
*Encyclopaedia of Islam*. New edition, ed. H.A.R. Gibb et al., Leiden: Brill, 1960–2003.
*Encyclopaedia of Islam*. New edition. Supplement, ed. C.E. Bosworth et al., Leiden: Brill, 1980–1997.
*Encyclopaedia of Islam*. Second edition, ed. Peri Bearman et al., published online.
Galtung, Johann. *Peace by Peaceful Means. Peace and Conflict, Development and Civilization*, Oslo/London: Sage Publications, 1996.
Al-Ghazali, Abū Ḥāmid, "The book of invocation," in: *Iḥyāʾ ʿulūm al-diīn*, translated by Kojiro Nakamura as *Ghazali on Prayer*, Tokyo: University of Tokyo, 1975.
Ibn Abī ʾl-Dunyā, *Aṣ-Ṣabr wa-ʾthl-thawāb ʿalayhi.*, Beirut: Dār Ibn Ḥazm, 1997.
Ibn al-ʿArabī, *An-Nāsikh waʾl-mansūkh fī al-qurʾān al-karīm*, Beirut: Dār al-kutub al-ʿilmiyya, 1997.
Ibn Ḥajar al-ʿAsqalānī, *Tahdhīb at-tahdhīb*, ed. Khalīl Maʾmūn Shīḥā et al., Beirut: Dār al-ma ʿrifa, 1996.
Ibn Hazm, *Kitab al-Fisal fi al-milal wa-ʾl-ahwaʾ wa-ʾl-nihal*, Cairo: Muʾassasat al-khanjī, 1903.
Ibn al-Jawzī, ʿAbd ar-Raḥmān, *Nawāsikh al-Qurʾān*, Beirut: Dār al-kutub al-ʿilmiyya, n.d.
Ibn Kathīr, *Tafsīr al-Qurʾān al-ʿaẓīm*, Beirut: Dār al-Jīl, 1990.
Ibn Khallikān, *Wafayāt al-aʿyān wa-anbāʾ abnāʾ al-zamān*, ed. Ihsān ʿAbbās, Beirut: Dār Ṣādir, n.d.
Ibn Qāḍī Shuhba, *Ṭabaqāt al-shāfiʿīyya*, Hyderabad: Dā'irat al-Maʿārif al-'Uthmāniyya, 1978.
Ibn Saʿd, Muḥammad, *Aṭ-Ṭabaqāt al-kubrā*, ed. Muḥammad ʿAbd al-Qādir ʿAtā, Beirut: Dār Ṣādir, 1998.

Johansen, Robert C., "Radical Islam and Nonviolence. A Case Study of Religious Empowerment and Constraint among Pashtuns," *Journal of Peace Research* 34 (1997), 53–71.
Khadduri, Majid, *War and Peace in the Law of Islam*, Baltimore: Johns Hopkins University Press, 1955.
Khan, Wahiduddin, *The True Jihād. The Concept of Peace, Tolerance and Non-Violence*, New Delhi: Goodword Books, 2002.
Mahmassani, Subhi, "The Principles of International Law in the Light of Islamic Doctrine," *Recueil des Cours* 117 (1966), 249–79.
Morabia, Alfred, *Le Ğihâd dans l'Islam medieval. Le "combat sacré" des origines au XIIe siècle*, Paris: Albin Michel, 1993.
Mottahedeh, Roy/al-Sayyid, Ridwan, "The Idea of the *Jihad* in Islam before the Crusades," in: Angeliki E. Laiou/Roy Parviz Mottahedeh (eds), *The Crusades from the Perspective of Byzantium and the Muslim World*, Washington, D. C.: 2001.
Mujāhid b. Jabr, *Tafsīr Mujāhid*, ed. ʿAbd ar-Rahmān aṭ-Tāhir b. Muhammad al-Surtā, Islamabad: Majmaʾ al-buḥūth al-islāmiyya, n.d.
Muqatil b. Sulayman, *Tafsīr*, ed. ʿAbd Allāh Maḥmūd Shiḥāta, Beirut: Muʾassaṣat at-taʾrikh al-ʾarab, 2002.
An-Nawawi, *Forty Ḥadīth*, trans. Ezzeddin Ibrahim and Denys Johnson-Davies, Cambridge, U.K.: Islamic Texts Society, 1997.
Omar, Irfan A., "Islam and the Other. The Ideal Vision of Mawlana Wahiduddin Khan," *Journal of Ecumenical Studies* 36 (1999): 423–39.
Al-Qurtụbī, Muḥammad. *Al-Jāmiʿ li-aḥkām al-qurʾān*. Beirut: Dār al-kitāb al-ʿarabī, 2001.
Ar-Rāzī, Fakhr ad-Dīn. *Al-Tafsīr al-kabīr*. Beirut: Dār iḥyāʾ al-turāth al-ʿarabī, 1999.
Renard, John. "*Al-Jihad al-Akbar*. Notes on a Theme in Islamic Spirituality," *Muslim World* 78 (1988), 225–42.
Saiʿīd, Jawdat, *Non-Violence. The Basis of Settling Disputes in Islam*, trans. Munzer A. Absi and H Hilwani, Damascus: Dār al-fikr, 2002.
Ash-Shāfiʿī, *Kitāb al-umm*, Bulaq: Maktaba al-kubrā al-amiriyya, 1903.
Ash-Shāfiʿī, *Al-Risāla*, ed. Aḥmad Shākir, n.pl., 1891.
Sezgin, Fuat, *Geschichte des arabischen Schrifttums*, 9 vols., Leiden: Brill, 1967–1984.
Ash-Shaybānī, *The Islamic Law of Nations. Shaybani's Siyar*, trans. and ed. Majid Khadduri, Baltimore: Johns Hopkins University Press, 1966.
Aṭ-Ṭabarī, Muḥammad b. Jarīr, *Jāmiʿ al-bayān fi tafsīr al-Qurʾān*, Beirut: Dār al-kutub al-ʾilmiyya, 1997.
Aṭ-Ṭabarī, *Ikhtilāf al-fuqahāʿ*, Cairo: n.p., 1933.
Aṭ-Ṭahāwī, Aḥmad, *Kitāb al-Mukhtaṣar*, ed. Abū al-Wafā al-Afghānī, Hyderabad: Lajnat iḥyāʾ al-maʿārif al-nuʿmāniyya, 1950.
Teichman, Jenny, *The Philosophy of War and Peace*, Exeter, UK: Imprint Academic, 2006.
Az-Zamakhsharī, *Al-Kashshāf ʿan haqāʾiq ghawāmiḍ at-tanzīl wa-ʿuyūn al-aqāwīl fi wujūh al-taʾwīl*, ed. ʿĀdil Aḥmad ʿAbd al-Mawjūd and ʿAl Muḥammad Muʿawwid, Riyadh: Maktabat al-ʿubaykān, 1998.
Az-Zuḥaylī, Wahba. *Āthār al-ḥarb fī ʾl-fiqh al-islāmī: dirāsa muqārana*, Damascus: Dār al-fikr, 1982.

# Suggestions for Further Reading


Abou El Fadl, *The Place of Tolerance in Islam*, Boston: Beacon Press, 2002.

Afsaruddin, Asma. *Jihad. What Everyone Needs to Know*, Oxford: Oxford University Press, 2021.

Bonney, Richard. *Jihad. From Qur'an to bin Laden*, New York: Palgrave Macmillan, 2004.

Al-Dawoody, Ahmed. *The Islamic Law of War. Justifications and Regulations*, New York: Palgrave Macmillan, 2011.

Esposito, John L. *Unholy War. Terror in the Name of Islam*, Oxford: Oxford University Press, 2002.

Hashmi, Sohail. *Just Wars, Holy Wars, and Jihads. Christian, Jewish, and Muslim Encounters and Exchanges*, Oxford: Oxford University Press, 2012.

Kelsay, John/Turner Johnson, James (eds), *Just War and Jihad. Historical and Theoretical Perspectives on War and Peace in Western and Islamic Traditions*. New York: Greenwood Press, 1991.

Paige, Glenn D./Satha-Anand, Chaiwat/Gilliatt, S. (eds), *Islam and Non-Violence*, Honolulu: University of Hawaii Press, 1993.

Sachedina, Abdulaziz A., "The Development of Jihad *in* Islamic Revelation and History," in: James Turner Johnson/John Kelsey (eds), *Cross, Crescent, and Sword. The Justification and Limitation of War in Western and Islamic Law*, 35–50, New York: Greenwood Press, 1990.

Smock, David and Qamar-ul Huda, *Islamic Peacemaking since 9/11*, Washington, D.C.: United States Institute of Peace, 2009.

Georges Tamer, Katja Thörner and Wenzel M. Widenka

# Epilogue

## Introduction

Through our inquiries into the concept of peace in the context of Judaism, Christianity and Islam we have attempted, at first, to strip the concept of its political and military dimensions and concentrate on those aspects dealing with the relationship between God and man. This is, in our view, the locus from which the duty and the will to engage in political peacebuilding emerge. It allows us to find appropriate theological resources from within the three faiths, which make it possible to develop a society in which everyone can live in peace and freedom, especially those whose sense of agency in the public sphere is religiously motivated. How does each one of these religions conceptualize peace? At which conceptual intersections do they meet? Which distinctive elements make their concepts of peace different from each other?

## 1 The Concept of Peace in Judaism

In Judaism, it is most appropriate to call the genuinely religious dimension of peace "messianic peace." The Hebrew word *shalom* defines a certain "fulfillment"; such a messianic peace is regarded as resolving all earthly conflicts. It also depicts the resolution of all conflicts between man and God.

One can identify three fundamental elements of messianic or prophetic peace in Judaism, which can be described most aptly in terms of "anti-politics," the "unity of opposites" and the "knowledge of God." The first element refers to an attitude in which the current means of political peacebuilding like negotiating, building alliances and strategical employment of power are replaced in their entirety by a more spiritual set of normative commitments emerging from a fundamental human awareness of God. What may seem, from a secular point of view, to be a fanciful idea has its roots in the prophets of the Hebrew Bible. They do not, in any sense, represent what one expects from a leader. They stammer and stutter while needing God's support in order to proclaim the message they are obliged to bear. They have no actual power. The same goes for the rabbis, whose interpretation of God's word in the scripture is regarded as equal to, or even surmounting prophecy. In addition, some central stories in the Hebrew Bible can be read as parables of "power" reflecting anti-political attitudes. Abraham

https://doi.org/10.1515/9783110682021-004

raises his offspring in the desert, as a shepherd, far away from the bustling cities. The efforts of the people of Babel to build their tower can be interpreted as an act of self-protection as well as an attempt to shape a human society that seals them off from the presence of God. The collapse of this enterprise demonstrates the sovereignty of God, which frustrates every plan built on mere human planning. "Anti-politics" does not constitute quietism, but rather a reminder of God's omnipotence along with an urgent appeal to internalize prophetic exhortations and turn back to God. The theological precedent for this idea lies in the concept of God's self-concealment through which he creates a sphere of freedom for his human creation. The most famous expression of this thought is found in Isaak Luria's concept of *tsimtsum*, or God's constriction, where God himself, originally filling the whole of reality, constricts himself in order to enable his creation to be free. Likewise, self-limitation is fundamental for the idea of anti-politics as it paves the way to a messianic era of peace capable of surmounting any current political action. It reflects a return to the paradisiacal state – an everlasting Sabbath.

This messianic era is characterized by the fulfillment of "the unity of opposites." The unity of everything which is divided and seems to be irreconcilable represents the reality of God himself along with his ultimate unity and oneness. Thus, this ideal state transcends time and secular reality. The creation returns to an everlasting Sabbath, where creation and creator, good and evil, and all presently incommensurable opposites are unified. In addition, the contradictions inherent in many Jewish laws and teachings will dissolve and form one great, peaceful vision. This indicates that ostensible theological contradictions in the present time point to a deeper unity which is still hidden to us and which will be revealed at a certain moment in the future.

Third, the "knowledge of God" is essential for understanding the concept of prophetic peace. It combines the two other elements – "anti-politics" and the "unity of the opposites" – towards a holistic vision of God's wholeness. *Knowing* here means submerging into God, becoming aware of everything that exists. This reflects a biblical conception of knowledge understood in terms of wholeness. The unity of all people, i.e. peace, is founded on the idea of the biblical covenant. Isaiah's vision of a child ruling once hostile opponents depicts a concept of a peace which reaches beyond the well-balanced and negotiated state of affairs apparent in the current world where wise or not-so-wise leaders shape the destiny of peoples.

Judaism, as such, is oriented towards *shalom*, this constituting the last word of the Talmud. *Shalom* is indeed a name of God, and peace represents the constant struggle for uniting that which had formerly been separated. This leads to a vision capable of enabling this abstract concept of prophetic peace to become a guidepost for the resolution of current disputes. This point is especially salient, as it cannot be denied that religions play a crucial role in framing most of the ongoing conflicts currently raging in the world. Prophetic peace should therefore be viewed in terms of a "workable mode of living together." If circumstances allow for the fundamental principles of anti-politics to emerge, defined in terms of the eventual unity of contradictory points of view, participants in a conflict can possess a mutual knowledge of God. This would bring together all religious and secular voices in a vision of unity.

## 2 The Concept of Peace in Christianity

The Holy Scriptures – the so-called Old and New Testaments – conceive of the concept of peace both in a negative sense in terms of the absence of war as well as in the positive sense of the experience of God's salvation (of the chosen people in the Old Testament) and reconciliation (through Jesus Christ in the New Testament) on earth. However, in this world, both aspects of peace rarely coexist and are never in a perfected state. Therefore, a human longing for peace always persists. Indeed, the Bible clearly states that both forms of peace depend on God. God's commandments serve to limit violence and his covenant with his chosen people is the fundament upon which it is possible to establish a safe and just society. The promise of positive peace on earth will be fulfilled through the coming of God's chosen messenger, the Messiah. Christians believe that Jesus of Nazareth is this Messiah who is therefore attributed with the title "Christ," which is the Greek translation of the word "Messiah."

After the brutal crucifixion and death of Jesus, it was, first of all, St. Paul who elaborated the assumption that the main aim of God's messenger, who was acclaimed by his followers the Son of God, was to overcome violence with love and reconciliation i.e. to bring to earth the peace of God. The resurrection of Christ demonstrates for believers that God can overcome death and transform injustice and brutality into glory. Thus, sin is understood in the New Testament essentially as a broken relationship between human beings and God, a state which is the origin of all evil deeds such as violence. However, since Jesus preaches that God is merciful and will forgive those who repent, it becomes possible for everyone to heal that relationship and overcome sin. This enables not only a new peaceful relation between God and mankind, but also a new state of peace between human beings. All religious controversies should come to an end and positive peace should flourish in Christian communities because they are united in Christ. Nevertheless, the hope that all quarrels will cease in Christian communities has not yet been fulfilled and so it becomes obvious that peace does not constitute a no-brainer, but rather must be understood as an ultimate but attainable goal arrived at through divine salvation.

The peacebuilding activities of Christians are not limited to their own congregations. For brotherly love, exemplified in Christ, should ideally spread into their social environment. Engagement on behalf of social welfare should become a special feature of Christian identity. Righteousness is a concept connecting redemption by God, the benevolent behavior towards one's neighbor and the task of establishing a just order.

The latter dimension touches particularly on the question of the relationship between spiritual and political power. Should Christians aspire to temporal power in order to create a peaceful society and potentially use violence in the attempt to achieve this goal? The New Testament demonstrates that Jesus always refused to take up weapons and did not even allow his disciples to do so. The peace of God is a spiritual peace which cannot be enforced by mundane power. Nevertheless, throughout history, Christians have had to wrestle with the question of how to deal with political power. Three fundamental conceptions can be differentiated in this context:

1) The conception of coexistence, according to which the mingling of mundane and spiritual affairs is strictly prohibited,
2) the apocalyptic conception of a contradiction between God and those secular forces deemed collectively as evil forces to be destroyed on Doomsday and
3) the conception of cooperation with the government. Such a government should be considered as installed by God himself, as long as it does not contradict God's will.

These conceptions have been often modified throughout the course of Church history and were challenged by historical events such as the rise of Christianity's political power with the Emperor Constantine or the Sack of Rome by the Visigoths in 410. This latter event prompted the Church Father St. Augustine to develop his influential distinction between God's kingdom as the City of God and the mundane kingdom as the City of Man. Since the City of God is considered to be otherworldly, eternal peace will therefore be realized only in heaven and not on earth. For Augustine, the City of God coexists in our times with the mundane City of Man. Augustine's aim was not to suggest that mundane power should not be utilized in order to support religious institutions, namely the Church, but to make clear that even a disastrous military defeat cannot disturb God's kingdom. Luther transformed this teaching from a relationship of coexistence to a relationship of cooperation between the two kingdoms.

The idea that mundane government has the duty to preserve peace – not eternal peace, but negative earthly peace – via political and even military means as an important condition to exercise religious duties and to preach the Gospel was in general highly accepted in the Christian tradition. However, in order to accept military interventions, it was necessary for certain rules and standards to be fulfilled. Therefore, debating the conditions for "just war" has constituted a prominent discussion in Christianity. Such a discussion is still going on among Christian theologians, especially while facing new forms of war like drone warfare.

After World War II, a shift from "just war theory" to the concept of "just peace" has taken place in Christian Churches all over the world. Crucial for this paradigm shift was a rejection of the link between war and justice, which has undergirded the tradition of "just war theory" a long time. War and violence can never be called just following this new paradigm. Furthermore, peace would need to be understood in terms of a general task, and not a mission solely concerned with armed conflicts. Peace should be conceived as peace within communities and marketplaces involving God's human creation. But even with this change of perspective, it remains a challenging task for Christians and the different churches to position themselves in regards to questions of humanitarian intervention and the duty to protect the suffering and the innocent, be it peacefully or by force.

From a religious perspective, humility and self-limitation are crucial to avoid growing intolerance and the proclamation of absolute und exclusive truth. These traits are essential for countering apocalyptic movements, theocracy and nationalism. Christians should be aware of religion's potential to create hatred as well as to produce peace. Peace will not be established among peoples and religions if there is no true dialogue between them along with a strong willingness to display humility and mutual acceptance. This demands a true interest in the other. It fundamentally calls for the exploration of one's own tradition as a basis for understanding and as a fundament for promoting peace. Thereby, the Churches themselves must become signs of peace while engaging in social ministry. With the proper display of humility, self-limitation and meekness, they may increase their credibility, particularly in times of crisis.

## 3 The Concept of Peace in Islam

Although peace/*salām* is very central in Islam and even one of the 99 names of God, Islamic thought has never been receptive to the idea of pacifism defined as the rejection of all forms of violence, especially as it concerns military intervention.

The impetus behind this phenomenon is not that Islam in any way constitutes a religion that promotes violence and aggressive behavior, but rather because it was confronted in a very early stage of its history with military realities. In addition, the possibility of passively acquiescing to injustice and violence contradicts the principle of *ḥisba* ("enjoining good and forbidding evil") according to which it is an individual and a collective moral responsibility to engage actively and steadfastly in the struggle against evil with the intention to achieve peace. This struggle is famously known as *jihād* which is often combined with the phrase "in the path of God" (*fī sabīl Allāh*). The Qur'ān understands the term as an individual or collective struggle to be achieved not only by military means. However, the meaning of the term has significantly changed throughout the centuries. Although the Muslims in the Meccan period were persecuted by the pagan Meccans, they did not make use of the right of self-defense. Accordingly, the Qur'ānic emphasis here rests upon an ethos of patiently bearing wrongdoings and injury. A prominent although presumably later verse in this regard is 3:200 which not only commands Muslims to "be patient and forbearing," but even to "vie in forbearance" while also imploring them to "be firm." But even when it came to fighting, the overall rules and principles of proportionality had to be respected. Qur'ānic ethics prohibited the commencing of hostilities and allowed for fighting mainly based on defensive purposes. The hermeneutical tool employed to nullify these Qur'ānic commands was that of abrogation (*naskh*) – a mechanism that has been debated in the field of Qur'ānic exegesis up to this day. However, it cannot be ignored that there exists a polyvalence of the term *jihād* within the Qur'ān itself. This has spawned a large variety of interpretations. The Abbasid period saw a form of secularization in the articulation of *jihād*, which had the effect of allowing expansionist wars. This tendency ultimately undermined the rich early discussions on *jihād* and divided the world into *dār al-islam* (House of Islam) and *dār al-ḥarb* (House of War) for the first time. The caliphs had to uphold a kind of "cold war," a state of permanent military awareness. Here it is worthy to mention that the concept of *dar al-ḥarb* finds no basis within the Qur'ān. It was not only the Hanafite school of law which denied the possibility of waging a just war against unbelievers. But rather the border situation with the Byzantine empire forged the concept of *jihād* into a kind of *realpolitik*.

While this type of interpretation in regards to warfare may have been prevalent, there have always been some exegetes like aṭ-Ṭabarī, for instance, who have focused on the aspect of patience as it relates to the concept of *jihād*, thereby building upon an idea prominently featured in the Qur'ān. To strive in the way of God, in this case, relates to a certain steadfastness in fulfilling religious obligations, while consistently fighting against evil desires and inclinations. Patience and forbearance constitute the necessary personal characteristics for not relenting in this struggle. For aṭ-Ṭabarī, this aspect in the concept of *jihād* – which he calls the "greater *jihād*" – is superior to the militant aspects of *jihād* – the "lesser *jihād*."

Additionally, the *ḥadīth* literature stresses the dimension of "patience" in the concept of *jihād*. It promotes its own genre of interpretation (*faḍāʾil aṣ-ṣabr*) which accentuates patience as a highly estimated virtue. To be patient is considered here in terms of a competition of the pious, which will in turn be rewarded by God.

Yet it would be misleading to treat the concept of peace in Islam as merely proceeding from the term "peace" and conceived as the opposite of war. The notion of peace, or in Arabic "*as-salām*" (along with its derivatives), is used in the Qurʾān to signify the relationship between the faithful and God. It represents the ultimate spiritual aim of human life and leads towards eternal peace as a paradisiacal state of being. Indeed, it begins in the here and now. There are several verses in the Qurʾān, along with passages in the *ḥadīth* literature, in which the faithful are asked to bring together multiple hearts, thereby practicing reconciliation. Reconciliation here represents the gateway to paradise. To establish peace, enemies have to be transformed into allies by the means of forgiveness, clemency and forbearance. Unity and brotherhood are central for the Muslim community, both in the here and now as well as in the hereafter.

Modern Muslim thinkers like Muḥammad ʿAbduh name this peaceful and patient attitude "love" (*al-ḥubb*) and extend the concept of reconciliation based on justice and love in order to include all human beings. He also stresses the peaceful meaning of *jihād* and refuses to reduce the concept of *jihād* to the commandment to fight against non-Muslims. ʿAbduh emphasizes the centrality of the peaceful aspects of Islam as opposed to its negative image propagated by many Orientalists and Western missionaries.

Several Muslim thinkers follow ʿAbduh's non-violent depiction of Islam and theologically undergird the idea of *jihād* with the notion of peaceful activism, using it in some cases as a means to push back against colonialist oppression. The contemporary Indian scholar Wahiduddin Khan refers to the portrayal of prophet Muḥammad, which is delivered in Islamic tradition and reveals in many aspects that the prophet always sought to find non-violent ways to solve conflicts or to reduce violent means in military conflicts to a lower limit in order to establish peace. For Khan, it is clear that "the position of peace in Islam is sacrosanct." Peaceful interactions between Muslims and non-Muslims are for him, last but not least, the basis of serious dialogue between them. This constitutes a form of intellectual stimulation which Muslims – and one should add: not only Muslims – are badly in need of "if they are to tread the path of progress."

# 4 Commonalities and Differences

The concept of peace can be deployed at least in three different modalities:

1) As the opposite of war in the sense of an armed conflict or a tense situation of mutual suspicion and aggression between states or rival groups. In this sense, peace can be defined as a relationship between states or groups, which excludes hostility and violence, enabling them to live in freedom from fear of violent conflicts.
2) In a more ethical sense, peace can be understood as a virtue defined in terms of an inner state of serenity and calmness. Or to put it the other way around: peace, in the sense of a peaceful state of mind, is characterized by the absence of inner struggles and false desires, thereby enabling an individual to have an autonomous life and behave in a deliberate and virtuous way.
3) Peace can also designate a kind of a utopic, everlasting state of affairs in which both aspects coincide and create a stable equilibrium all over the world – and/or in the afterworld. It is characterized by the absence of actual inner and outer conflicts and even by the absence of any threat of potential conflicts.

All three meanings are present in the Sacred Scriptures of Judaism, Christianity and Islam and in their theological traditions. However, these different meanings are emphasized in various fashions dependent on religious trends and changing political and social circumstances. In the face of current violent conflicts between groups that justify violence with religious reasons, the role of peacemaking – in the sense of promoting "peace" in the political meaning – dominates in debates about religion and peace. Thus, the prevailing question is: Which impact may religions have on their adherents so that they can live together without hostility and violent conflict? To refer to individual virtues or ideal states of everlasting harmony would obviously fall short of the mark if the aim is to give fruitful answers to the question of how to solve religiously motivated conflicts.

When we consider the use of the word "peace" and its equivalents in Hebrew, Greek and Arabic in the Sacred Scriptures, it becomes obvious that in a religious context the second and third meanings of peace should prevail. Alick Isaak has noted that not only in Judaism, but also in Christianity and Islam, peace constitutes both a "key value" and a "central organizing principle." It represents, in fact, the ultimate aim of human existence and behavior. Yet first of all, the concept of peace in Judaism, Christianity and Islam must designate the relation between human beings and God.

Regarding the first field of semantic meaning associated with peace (as articulated above), a crucial question, in past and present, is related to Christian communities equipped with political power, while the struggle for peace has been understood there as either a political enterprise or a spiritual challenge. We can find a huge number of answers to this dilemma ranging from pacifism and quietism to the advocating of military interventions. Here we can observe a parallel situation to the understanding of *jihād* or the "striving in the path of God" in Islam. However, while Christianity presents a sharp distinction between the realm of mundane power and the realm of spiritual power based on the New Testament, the term *jihād* in the Qur'ān is characterized by a high degree of polyvalence. Consequently, interpretations of this notion have varied across the ages, from spiritual conceptualizations that have understood *jihād* as a steadfast and patient inner struggle to overcome evil inclinations to those which promote a practical, more militaristic duty. Although the Qur'ān and the Bible do not explicitly advocate waging "holy wars," one can find this idea in both traditions in connection with expansionary and missionary contexts. Yet all three traditions contain deliberations regarding the extent to which military means can be seen as adequate, i.e. the so called "just war theories."[1]

A central corrective against violent tendencies within the three faiths is the fact that all of them attribute God with peaceful qualities and intentions. God is indeed characterized as merciful, with numerous passages in the Sacred Scriptures of Judaism, Christianity and Islam explaining that God loves and rewards those who show forbearance to wrongdoers. God is characterized as the (only) source of inner peace, with all three traditions developing practices and theories regarding the way how to reach this inner form of spiritual peace. Furthermore, in all three traditions peace depends on God. All human efforts to establish peace on earth are in vain if they are not established on the foundation of faith in God.

But these exhortations should not be understood as appeals to quietism. In all three traditions, it is clearly stated that earthly peace becomes a reality only through the struggle for peace and the avoidance of violence by man. Biblical texts as well as the Qur'ān not only revere the commandment to treat thy brother with love and mercy, but also implore mercy towards oppressors and enemies. Here the concept of peace is closely related to the idea of justice and the equality of men. Peace can only flourish in a just society whereby everyone treats his or her neighbor in the manner that they wish themselves to be treated. In opposition to secular formulations of this maxim such as Immanuel

---

1 See volume 18, „The Concept of Just War in Judaism, Christianity and Islam" of the present book series (forthcoming).

Kant's categorical imperative, peacebuilding constitutes, according to all three religions (and particularly Islam), not merely altruistic forms of behavior. The award for the attainment of peace can be obtained in the afterlife. Judaism also advances the notion that the realization of peace on earth will accelerate the coming of the messianic age. In Christianity, peace does not constitute an end in itself, but is rather the result of redemption obtained through the incarnation and the death of Jesus Christ. Therefore, the concept of peace in all three religions points ultimately to the afterlife, where the ideal of peace will be fulfilled.

For Jews, Christians and Muslims, the age of eternal peace will begin with the coming of the Messiah / Christ and conclude with the last judgement as the fulfillment of ultimate justice. A special feature of Christian faith lies in the idea that the Messiah has already come in the person of Jesus Christ. With his incarnation, the power to establish peace has arrived into the world. The sending of the Messiah was an expression of God's love to human beings, which was then perfected through his ultimately redemptive death on the cross and subsequent resurrection. Therefore, in Christian thought, a kind of simultaneity exists between earthly attempts to build peace and a sphere of eternal peace in the process of actualization achieved through coming of Christ. However, although the coming of Jesus Christ can be understood as the dawn of an era of eternal peace, the lack of peace in the sense of the absence of hostility and violence remains a common experience for Christians from the very beginnings of Christianity up till now. After his resurrection, Christ returned to his disciples saying, "peace be with you" in order to take away the fear, sorrow and despair that have overwhelmed the faithful after the arrest and death of their "Lord." This salutation not only granted comfort and hope, but it also formed the basis upon which they were able to follow Jesus in his mission to bring the peace of God to humanity.

Facing recent and present social conflicts, abhorrent cruelty and destructive wars in several parts of the world, reconciliation has become a central motif in the process of peacebuilding. Jews, Christians and Muslims should increasingly strive to establish themselves as pivotal actors in this process whenever and wherever it is needed. Peacebuilding, however, can only be a pursued through patient dialogue.

# List of Contributors and Editors

**Asma Afsaruddin** is Professor of Near Eastern Languages and Cultures in the Hamilton Lugar School of Global and International Studies at Indiana University, Bloomington. She received her doctorate in Arabic and Islamic Studies from Johns Hopkins University in Baltimore, Maryland. She is the author and editor of eight books, including *Jihad: What Everyone Needs to Know* (Oxford University Press, forthcoming); the award-winning *Striving in the Path of God. Jihad and Martyrdom in Islamic Thought* (Oxford University Press, 2013) which is being translated into Bahasa Indonesian; and *The First Muslims. History and Memory* (OneWorld Publications 2008), which has been translated into Turkish and Bahasa Malay. Her fields of specialization include pre-modern and modern Islamic religious and political thought, Qur'ān and ḥadīth, and Islamic intellectual history.

**Alick Isaacs,** born in Scotland and emigrated to Israel in 1986; studied Jewish History, Literature and Philosophy at the Hebrew University in Jerusalem; obtained PhD from the Hebrew University in 2003. He is Co-Founder and Co-Director of Siach Shalom (Talking Peace) at Mishkenot Sha'ananim in Jerusalem and adjunct lecturer at Hebrew University. Publications include *A Prophetic Peace. Judaism Religion and Politics* (Indiana University Press 2011 [in Hebrew translation Shocken Books 2014]); "Shlomzion" published by Duke University Press in *Common Knowledge*, 20:1 Winter 2013/14; with Avinoam Rosenak and Sharon Leshem-Zinger, "Human Rights. On the Political, the Dynamic and the Doctrine of the Unity of Opposites," in: *The Role of Religion in Human Rights Discourse* edited by Hanoch Dagan, Yedidia Z. Stern, and Shahar Lifshitz (Israel Democracy Institute, 2014). Current areas of research, Peace in Jewish Thought and Western Thought, Heidegger's Ontology and its Implications for reading Jewish texts.

**Volker Stümke** studied Theology and Philosophy in Hamburg, Tübingen and München. He is the Leading Academic Director of the Institute of Ethics at the Führungsakademie der Bundeswehr Hamburg and Professor of Systematic Theology at the University of Rostock. Publications include: *Das Friedensverständnis Martin Luthers. Grundlagen und Anwendungsbereiche seiner politischen Ethik* (Kohlhammer, 2007), *Zwischen gut und böse. Impulse lutherischer Sozialethik* (LIT Verlag, 2011), "Religion und Gewalt. Eine Rezension" (in: ThR 84, 2019, 40–95; 105–157).

**Georges Tamer** holds the Chair of Oriental Philology and Islamic Studies and is founding director of the Research Unit "Key Concept in Interreligious Discourses" and speaker of the Centre for Euro-Oriental Studies at the Friedrich-Alexander-University of Erlangen-Nuremberg. He received his PhD in Philosophy from the Free University Berlin in 2000 and completed his habilitation in Islamic Studies in Erlangen in 2007. His research focuses on Qurʾānic hermeneutics, philosophy in the Islamic world, Arabic literature and interreligious discourses. His Publications include: *Zeit und Gott. Hellenistische Zeitvorstellungen in der altarabischen Dichtung und im Koran* (De Gruyter, 2008), and the edited volumes *Islam and Rationality. The Impact of al-Ghazālī* (Brill, 2015) and *Exegetical Crossroads. Understanding Scripture in Judaism, Christianity and Islam in the Pre-Modern Orient* (De Gruyter, 2017).

https://doi.org/10.1515/9783110682021-005

**Katja Thörner** is research assistant in the Research Unit "Key Concept in Interreligious Discourses" at the Friedrich-Alexander-University of Erlangen-Nuremberg. She studied Philosophy and German Literature in Trier, Würzburg and Berlin and received her PhD in Philosophy at the Munich School of Philosophy in 2010. She is author of *William James' Konzept eines vernünftigen Glaubens auf der Basis religiöser Erfahrung* (Kohlhammer, 2011) and published with Martin Turner, *Religion, Konfessionslosigkeit und Atheismus* (Herder, 2016) and in collaboration with Trutz Rendtorff, *Ernst Troeltsch. Schriften zur Religionswissenschaft und Ethik (1903–1912),* (De Gruyter, 2014). Her research focuses on the philosophy of religion, theories of interreligious dialogue and comparative studies of concepts of the hereafter in Islam and Christianity.

**Wenzel Maximilian Widenka** studied History, Catholic Theology and Interreligious Studies at the Universities of Bamberg and Vienna. He received his PhD in Jewish Studies at the University of Bamberg in 2019 with a study about the struggle for religious emancipation of $19^{th}$ century Jews on the countryside. He is currently working as a research assistant for the interdisciplinary project „Key Concepts in Interreligious Discourses," as well as for the Chair of Economic and Social History of the Catholic University of Eichstätt-Ingolstadt. Recent publication: *„Sehet, da kommen Schakale, den Weinberg zu zerstören, den Weinberg Israels." Emanzipation und Konfessionalisierung im fränkischen Landjudentum in der ersten Hälfte des 19. Jahrhunderts"* (University of Bamberg Press, 2019).

# Index of Persons

Aaron 3, 70
Abba Sikra 18
ʿAbd Allāh b. ʿUmar 102 n6, 106, 113, 133
ʿAbd al-ʿAzīz b. Abī Rawwād 130
ʿAbd al-Ghaffār Khān 146
ʿAbd ar-Razzāq 119, 127, 127 n80, 129, 129 n86
ʿAbda b. Abī Lubāba 142
Abdimi of Haifa 17
ʿAbduh, Muḥammad 142–143 n117, 144 n123, 146, 154
Abel 10, 46
Abraham 5, 11, 11 n32, 12, 29, 32, 47, 56, 56 n35, 159
Absalom 12
Abū Bakr 136, 137
Abū Hurayra 104, 104 n15, 134
Abū Jahl 136
Abū Lahab 136
Abū Mujliz 127, 128
Abū Nuʿaym al-Ḥāfiẓ 106
Abū Rawq 116
Abū Salamah 104, 104 n15, 106
Abū Sufyān b. Ḥarb 137
Abū ʿUmar 106, 106 n19
Achaz 29
Adam 9, 13, 31, 35, 146, 147, 148, 149
Ahasuerus 40 n98
ʿĀʾisha 104 n15, 134, 150
Amemar 17, 18
Amitai 15
Amon 29
Amos 13, 13 n44
Apollos 51 n15
Ashlag, Yehuda 21
ash-Shāfiʿī 124, 124 n66, 124 n67
Augustine 64, 65, 66, 67, 76, 162
al-Awzāʿī 113, 122

Bar Kappara 28
Bartolomé de Las Casas 68 n65, 76
Ben Arach, Eliezer 25
Ben Chalafta, Shimon 36
Ben Samuel, Judah 40 n98
Ben Zakkai, Yohanan 18, 19, 20, 24, 34
Benedict XVI, Pope 89
Bonhoeffer, Dietrich 84 n96
Breslav, Nachman of 20
Buber, Martin 31, 32, 32 n79, 32 n80
al-Bukhārī 130, 133, 150

Caesar 60, 64 n52
Cain 10, 46, 47, 65
Chernobyl, Rabbi Menahem Mendel of 38 n91
Chizkiah 29
Cicero 67
Constantine 63, 64
Cornelius 63

Daniel 48 n9, 60
David 12, 24
Derrida, Jacques 10 n26

Elijah 12
Esaias 49 n12
Esau 12
Esther 40 n98
Eusebius 63
Eve 9, 13, 31
Ezekiel 13

Fisch, Menachem 15 n55, 19 n59, 27 n72
Francisco de Vitoria 76, 76 n82

Gabriel 28
Gamliel, Rabban 37
Gopin, Marc 2 n5
Green, Arthur 22, 23, 23 n65

Halbertal, Moshe 15 n55
Halevi, Judah 38 n90
Haran 12
Hartman, David 15 n55
al-Ḥasan al-Baṣrī 102, 119, 127, 137
Heschel, Abraham J. 10 n23, 14, 14 n50
Hirsch, Samson Raphael 9, 9 n19
Hishām b. Ḥakīm b. Ḥizām 117

https://doi.org/10.1515/9783110682021-006

Hobbes, Thomas 82, 82 n90
al-Hudaybiyya 110, 151, 153

Ibn ʿAbbās 102 n6, 107 n22, 109, 119, 136, 137
Ibn Abī ʾl-Dunyā 126, 126 n74, 127, 127 n77, 128, 128 n82, 130, 130 n87
Ibn ʿAṭiyya 114
Ibn Hanbal, Aḥmad 129
Ibn Hazm 130
Ibn Isḥāq 139
Ibn Jurayj 103, 113
Ibn Kathīr 115, 115 n43, 138, 138 n105, 140, 140 n113, 141
Ibn Māja 127
Ibn Muḥakkam 112, 119
Ibn Saʿd 104 n12, 104 n15
Ibn Zayd 120
Ikrima 111
Isaac 5, 11, 12
Isaacs, Alick 1, 166
Isaiah 5, 13, 14, 21, 160
Ishmael 12
Iṣma Abī Ḥukayma 126

Jābir b.ʿ Abd Allāh 104
Jacob 5, 11, 12, 49 n10
Jawdat Saʿīd 146–149
Jeremiah 13
Jesse 12, 32, 47
Jesus 46, 49, 50, 51, 51 n15, 51 n16, 52, 52 n20, 53, 53 n21, 53 n23, 54, 55, 55 n29, 56, 56 n34, 56 n35, 57, 58, 58 n41, 59, 59 n43, 60, 61, 64, 65, 69, 70, 71, 73, 74, 75, 75 n76, 76, 80, 81, 87, 88, 90, 161, 162, 168
John the Baptist 63 n49
Jonah 15
Joseph 12, 24
Joshua 64 n53
Judah 14, 24
Judah Loew of Prague 27

al-Kalbī 119
Kaminsky, Howard Gary 2 n5, 3 n7
Kant, Immanuel 77, 77 n84, 167
Kaplan, Aryeh 8 n15
Kaufman, Tsippi 38 n95
Khadduri, Majid 124 n68, 124 n69, 124 n70
Kimball, Charles 91, 92, 92 n117
Konrad, George 7, 8 n14
Kook, Abraham Isaac 6 n12, 9, 9 n21, 20 n61, 27, 28, 28 n74, 31, 35, 35 n82, 37, 37 n87, 38, 38 n92, 38 n93, 39, 39 n96, 41
Korach 39, 39 n97
Küng, Hans 93, 93 n121

Laidi, Shneur Zalman of 34 n81
Leshem Zinger, Sharon 1, 1 n2, 1 n4, 27, 28 n75, 41, 42
Levi 36
Loew ben Bezalel, Judah ("Maharal of Prague") 27, 31, 38
Lot 11, 12, 40 n98
Luke 49, 49 n12, 50, 56, 57, 58, 58 n39, 58 n40, 59, 60, 61, 63, 63 n49, 69, 76
Luria, Isaak 160
Luther, Martin 65, 66, 66 n57, 68, 69 n66, 71, 72, 73, 74, 75, 75 n79, 76, 76 n80, 80, 80 n87, 80 n88, 82 n91, 92, 162

Maimonides 31, 37 n89
Matthew 59, 60, 61, 70
Mawdudi 150
Michael, Angel 28, 39 n97, 48 n9
Mordechai 29
Moses 12, 13, 14, 22, 47, 70, 148
Muhammad 101, 103 n8, 104 n15, 106, 107, 108, 112, 113, 115, 116, 117, 120, 134, 136, 137, 139, 141, 144, 147, 148, 150
Muqātil b. Ḥayyān 119
Muqātil b. Sulaymān 103, 110, 111, 112, 115, 119, 136
Muslim b. al-Ḥajjāj 117, 127, 134

Nachman of Breslav 20, 20 n61
Neher, André 28 n74, 38 n95

Paul, Apostle 50, 51, 51 n15, 52, 53, 58, 61, 70, 72, 75, 80, 80 n89, 161
Peter, Apostle 58, 60

Philip, Apostle 53 n22
Pinker, Steven 83, 83 n92

Qatāda b. Di'āma 103, 136
al-Quraẓī 103 n10, 106
al-Qurṭubī 106, 107, 117

Rabbi Meir 25, 26, 27
Rabbi Nehemiah 25
Rabbi Yoshiah 29
ar-Rabīʿ 111
ar-Rāzī 104, 105, 105 n16, 106, 111, 113, 114, 116, 121, 121 n59, 137, 138 n104
Reuben 12, 28 n74
Ridwan as-Sayyid 122, 122 n60
Roness, Michal 2 n5
Rosenak, Avinoam 1, 1 n2, 1 n3, 6 n12, 20 n61, 28 n74, 37 n87, 37 n88, 38 n93, 38 n95, 41
Roth, Daniel 2 n5, 3 n8
Roy Mottahedeh 122, 122 n60

Sagi, Avi 16 n55
Sa'd b. Abī Waqqās 147
Ṣafwān b. Salīm 117
Senghaas, Dieter 54, 54 n27, 85 n100
ash-Shāfiʿī 124, 124 n66, 124 n67
Shimon Ben Chalafta 36
Shmuel 27
Shuraḥbīl b. Saʿīd 104, 104 n12
Simon, Apostle 59
Solomon 12
Soloveitchik, Joseph Dov 5 n10
Sommer, Benjamin 14 n51
Steinberg, Gerald M. 2 n5
as-Suddī 110, 111, 119, 136
Sufyān ath-Thawrī 114, 122
Sufyān b. 'Uyayna 104 n12, 127

Sutton, Avraham 21 n62
as-Suyūṭī Syed 143

aṭ-Ṭabarī 111, 112, 113, 115, 116, 119, 120, 121, 123 n62, 123 n63, 125, 136, 137, 138, 139, 141, 164
aṭ-Ṭahāwī, Aḥmad 124 n70
Terach 29
Theodosius 63
at-Tirmidhi 129
Thomas Aquinas 67, 67 n61, 78, 78 n86, 79
Tishby, Isaiah 38 n90
Tsadok Hakohen of Lublin 40 n98

Ubayy b. Ka'b 102 n6
'Umar 103 n8, 104 n15
'Umayr b. Sa'd 117
'Urwa b. az-Zubayr 104 n15
'Uthmān b. 'Affān, Ṭalḥa 104 n15
Umm Salama 104 n15

Vespasian, Titus Flavius 19, 34

Wahiduddin Khan Mūsā b.ʿUqba 146, 150, 150 n148, 153, 165
al-Wāqidī 104 n12
Weingardt, Markus A. 90, 91 n115
Weinreb, Friedrich 9 n21, 10 n24, 11 n30, 22 n64

Yohanan Ben Zakkai 18, 19, 20, 24, 34
Yoshiah 29

az-Zajjāj 138
az-Zamakhsharī 111, 115, 137, 137 n103
Zinger, Sharon Leshem 1, 1 n2, 1 n4, 27, 28 n75, 41, 42
az-Zuhrī 104 n15

# Index of Subjects

Abbasid 116, 122, 125, 164
Ancient Church 51, 55

Babel 10, 11, 160
Battle of Badr 113
brotherly love 46, 55, 141, 162
– neighborly love 53, 55, 72, 73, 74

conscience 15, 61, 72, 74, 75, 76, 77, 78, 79

*da'at Hashem* 31, 32, 33, 34, 35

Egypt 5, 11, 12, 142
Europe 7, 76, 82, 142, 144
exile 3, 4, 8, 23, 25, 47

*faḍā'il al-jihād* 126
*faḍā'il aṣ-ṣabr* 126, 127, 165
*fī sabīl allāh* 100, 103, 129, 164
freedom 7, 8, 54, 69, 74, 75, 76, 77, 79, 80, 81, 82, 91, 159, 160, 166

government 7, 59, 65, 66, 67, 68, 69, 70, 71, 72, 77, 78, 79, 82, 86, 87, 88, 89, 92, 122, 124, 148, 151, 162

Ḥasidism 40
– Ḥasidic 34, 38, 40
hermeneutics 15, 16
*Hester Panim* 8, 20, 21, 22
Holy War 93, 107, 121, 123, 153, 167

Israel 1, 3, 4, 5, 6, 7, 8, 11, 12, 15, 22, 23, 29, 30, 32, 36, 37, 38, 40, 41, 42, 46, 47, 57, 64

*jāhiliyya* 140, 145
Jerusalem 1, 11, 14, 16, 17, 18, 19, 20, 25, 36, 37, 46, 49, 64
*jihād* 100–118, 120, 121–126, 127, 128, 129, 130, 131, 142, 143, 144, 146, 147, 148, 149, 150, 151, 152, 153, 154, 164, 165, 167
Judgement 58, 80, 81, 168
just war 64, 67, 68, 82, 83, 84, 163, 164, 167
justice 33, 45, 47, 54, 66, 81, 84, 85, 86, 90, 99, 105, 109, 132, 133, 134, 145, 147, 148, 149, 151, 153, 163, 165, 167, 168
justification 4, 57, 70, 72, 81, 121

Kabbala
– Kabbalist 8, 21, 37, 38
kingdom 14, 23, 51, 52, 58, 59, 60, 64, 65, 66, 162

Mecca 101–107, 108, 109, 110, 111, 112, 119, 122, 141, 143, 145, 150, 151, 164
Medina 101, 102, 103, 104, 108–110, 113, 114, 122, 139, 140, 141, 144, 145, 151, 154
Messiah 37, 47, 48, 49, 50, 55, 57, 161, 168
Middle Ages 37, 68, 72
– Medieval Ages 69
Middle East 2, 5, 6, 7, 37, 40, 42
Midrash 10, 19, 20, 24, 29, 30
*mujāhid* 150

*naskh* 121, 122, 143, 164

Orthodox Churches 70

Palestine 117
– Palestinians 1, 41
paradise 132, 134, 136, 137, 165
– Garden of Eden 9, 11, 22
Peace Churches 73, 83, 87, 88
pluralism 24, 27, 39, 93
prophecy 12, 14, 16, 17, 18, 20, 21, 25, 32, 33, 34, 41, 105, 159
prophet(s) 3, 4, 8, 13, 14, 15, 16, 17, 18, 24, 33, 38, 49, 53, 56, 60, 100, 101, 103, 104, 106, 108, 109, 112, 113, 114, 117, 120, 121, 127, 128, 129, 130, 133, 134, 136, 137, 139, 140, 141, 145, 147, 148, 150, 151, 152, 153, 159, 165
Protestant Churches 74, 76, 88

*qitāl* 100, 108, 122, 150, 151

https://doi.org/10.1515/9783110682021-007

reconciliation 3, 13, 50, 52, 54, 55, 56, 57, 86, 90, 124, 135, 138, 139, 141, 144, 145, 152, 154, 161, 165, 168
Reformation 72, 148, 150
Roman Catholic Church 70
– Catholic Church 70, 73, 76
Rome 19, 64, 162

Sabbath 10, 11, 24, 39, 49, 160
*salām* 99, 163
*shahīd* 127
*shalom* 1, 3, 4, 20, 29, 31, 37, 159, 160
*spiritualia* 70, 71, 89
sword 48, 49, 52, 59, 61, 64, 66, 69, 70, 71, 72, 73, 74, 83, 115, 118, 123, 143
*ṣabr* 105, 106, 107

*tafsīr* 103, 110, 112, 115, 116, 119, 121, 140
temple 11, 17, 18, 19, 20, 25, 29, 47
*temporalia* 70, 77, 82
Tree of Life 11, 13, 15, 31
*tsimtsum* 8, 22, 160

Umayyads 114, 116, 122, 125

war 10, 11, 14, 31, 45, 46, 47, 49, 64, 65, 66, 67, 68, 72, 74, 76, 81, 82, 83, 84, 85, 86, 87, 88, 91, 92, 99, 100, 108, 109, 114, 116, 118, 119, 121, 122, 123, 124, 131, 139, 140, 144, 150, 151, 152, 153, 154, 161, 163, 164, 165, 166, 167, 168

www.ingramcontent.com/pod-product-compliance
Lightning Source LLC
Chambersburg PA
CBHW051526230426
43668CB00012B/1750
* 9 7 8 3 1 1 0 6 8 1 9 3 2 *